Carbon Dioxide for Pool pH Control

A Practical Guide for Trained Professionals

David Watson

Published by: David Watson Pool Consulting
info@davidwatsonpoolconsulting.com
www.davidwatsonpoolconsulting.com

Design and illustrations: Sue Knox
Cover Design: David Watson

Ingram Spark / Amazon KDP Print-on-Demand Edition
ISBN EBook: 978-1-7645296-4-8
ISBN Paperback: 978-1-7645296-6-2
Printed in 2026

In preparing this handbook, AI tools have assisted with editing and verifying references across multiple international databases.

All content has been carefully reviewed and edited for accuracy and reliability, informed by my four decades of field experience and current regulatory standards.

Acknowledgement

I extend my sincere thanks to the early readers who reviewed the initial drafts of this book.

Your thoughtful feedback, careful reading, and constructive suggestions made this work stronger and clearer.

I am grateful for your support in developing this manuscript.

Contents

CHAPTER I
What This is About . . . 15

CHAPTER 2
Before You Read This Book . . . 17

CHAPTER 3
Larger Pools Are More Forgiving . . . 19

CHAPTER 4
How to Read the Chemistry in This Book . . . 21

PART 1 — FOUNDATIONS . . . 23

CHAPTER 5
What This Part Covers . . . 25

CHAPTER 6
Origins . . . 27

CHAPTER 7
Reader Skill Level and Scope of This Guide . . . 31

CHAPTER 8
Essential Safety Principles When Working with CO_2 . . . 33

CHAPTER 9
Folklore . . . 35

CHAPTER 10
When CO_2 Is Less Efficient . . . 37

CHAPTER 11
When CO_2 is More Efficient . . . 41

CHAPTER 12
Runaway Total Alkalinity . . . 43

CHAPTER 13
Expensive to Manage . . . 47

PART 2 — INSTALLING A CO_2 SYSTEM . . . 49

CHAPTER 14
How to Use the Install Chapters . . . 51

CHAPTER 15
Before Starting Installation 53

CHAPTER 16
Storage 55

CHAPTER 17
Sensors and Alarms 61

CHAPTER 18
Auto Changeover
and Regulator 65

CHAPTER 19
Tube and Fittings 69

CHAPTER 20
Protection 79

CHAPTER 21
Carbon Dioxide Gas Controller 81

CHAPTER 22
Injection Point 83

CHAPTER 23
Installation Complete 91

PART 3 — OPERATING A CO_2 POOL 93

CHAPTER 24
Day-to-Day Operation in Plain English 95

CHAPTER 25
pH Control 97

CHAPTER 26
Correcting Total Alkalinity 101

CHAPTER 27
How to Dose Acids 113

CHAPTER 28
Minimising Consumption 115

CHAPTER 29
CO2 Consumption Calculator 121

CHAPTER 30
CO_2 pool field troubleshooting 127

CHAPTER 31
Glossary 131

PART 4 — UNDERSTANDING THE SCIENCE. . . . 133

CHAPTER 32
Why the Science Matters (But Stays Practical) 135

CHAPTER 33
Appendix: Langelier Saturation Index 137

CHAPTER 34
Appendix: Henrys Law 143

Henry's Law Meets Langelier Saturation Index 149

PART 5 —TECHNICAL APPENDICES & TOOLS . . . 153

CHAPTER 36
Quick Reference and Extras 155

CHAPTER 37
Appendix: Calculating Tube Size Example 157

CHAPTER 38
Appendix: Painting Plastics 167

CHAPTER 39
Appendix: Installation and Commissioning Checklist . . . 169

CHAPTER 40
Appendix: Start Up Steps 171

CHAPTER 41
Appendix: Global Standards Crosswalk for CO_2 pH Control . 175

CHAPTER 42
About the Author 177

A pool is not just water.
It is a series of interrelated systems.

It is our responsibility.
We own every drop until it's safe.
When a bather walks out the gate,
they carry our standard with them.
They take it home, to work,
to their communities.

Their safety —
and the safety of everyone they touch —
rests with us.

CHAPTER I

What This is About

Real-world practical guidance

Have you ever been confronted with a carbon dioxide pH-controlled pool with:

- Total alkalinity that keeps increasing day after day, week after week, and it is blamed on carbon dioxide.
- pH that seems stable, but every week it is consuming more gas.
- Water that is grey-toned, looks lifeless and dull.

This book explores these issues and gives you real-world practical guidance on how to correct them.

Specifically, it covers:

- Suitability of a water body for CO_2 as a pH donor
- Causes of runaway total alkalinity
- Installation from storage to diffuser
- Ongoing operation of a CO_2-dosed water body

Some theoretical context is included to support the practical guidance. It does not need to be memorised, it is there to explain the why behind the issues.

This book explains why pH behaves the way it does. The companion ORP/pH book explains how controllers respond to that behaviour.

CHAPTER 2

Before You Read This Book

What You Should Already Know

This guide assumes familiarity with the following concepts. If any are unfamiliar, take time to review them before working through the chapters. The material will make significantly more sense with these foundations in place.

Basic Water Chemistry

- What pH means, how it is measured, and what the typical pool target range is
- What Total Alkalinity (TA) is and why it matters for pH stability
- What Calcium Hardness is and its role in water balance
- What Cyanuric Acid (stabiliser) does and how it affects chlorine activity
- What Total Dissolved Solids (TDS) represent

Chlorine Fundamentals

- The difference between free chlorine and total chlorine
- What sodium hypochlorite is and how it is dosed
- Why chlorine demand changes with bather load, temperature, and UV exposure
- The relationship between pH and chlorine effectiveness

Chemical Handling Basics

- How to read a Safety Data Sheet (SDS)
- What specific gravity means and how it relates to liquid chemicals
- Safe handling procedures for hydrochloric acid and similar chemicals
- What PPE is required when handling pool chemicals

Pool Systems and Equipment

- How a recirculation system works: pump, filter, return, balance tank
- What a dosing pump is and how it is used
- What a chemical controller does and how pH and chlorine probes function
- The difference between the suction and discharge sides of a pump

Basic Maths

- Calculating pool volume in litres
- Multiplying and dividing with decimals
- Reading and interpreting a simple table

CHAPTER 3

Larger Pools Are More Forgiving

Big pools buffer mistakes; small pools amplify them

One important misconception needs clearing before we go any further into CO_2.

There is a persistent view in the industry that larger water bodies are harder to manage than smaller ones. That is not correct, and a simple story illustrates why.

Ever killed a goldfish?

As a child, you probably visited a shop, picked out a small bowl and a goldfish, and brought them both home. Excited, you filled the bowl, released your new pet, and because no one told you a fish's stomach is roughly the size of its eye,loaded in what seemed like a reasonable amount of food.

Over the next day or two, the goldfish ate what it could. The rest sank, broke down, and clouded the water. By day two or three, the fish was floating upside down.

For many people, that was their first experience of death — and, more relevant here, their first proof that smaller volumes are harder to manage.

If you had put that same goldfish into a bathtub and fed it the same amount, it would very likely have survived. The difference is simply dilution.

The same principle applies to pools.

Imagine a 2,000,000-litre pool where a dosing error results in 100 kg of chemical being added instead of the correct 10 kg. That 90 kg overdose across 2,000,000 litres adds approximately 45 mg/L to the parameter in question. Annoying, and it will take time and product to correct, but manageable.

Now consider a portable spa of around 1,000 litres. The same proportional error 1 kg instead of 0.1 kg, adds approximately 900 mg/L. That is not a chemistry problem to work through. That is a drain-and-refill situation.

Large bodies of water dilute dosing mistakes. Small bodies concentrate them.

Larger water bodies are often perceived as more difficult to manage for reasons that have little to do with volume itself:

- High aeration features designed to impress rather than function efficiently
- Shallow interactive water zones that dominate chemical demand
- High bather loads relative to volume
- Strong wind exposure resulting from architectural design
- Regulatory targets that impose specific chemical levels
- Operational expectations carried over from smaller water bodies

When a larger water body is approached as a system, these challenges become manageable. CO_2 is most commonly used on larger water bodies, typically 150,000 litres or more, though not always at the 2,000,000-litre scale.

CHAPTER 4

How to Read the Chemistry in This Book

Chemical formulas are included throughout this book to give a complete picture

Readers do not need to memorise them. They are background context of the full story behind the practical guidance. Where possible, every equation is also written out in plain language.

Formula	Common Name	What it does in this book
HOCl	Hypochlorous acid	The active, fast-acting sanitising form of chlorine
NaOCl	Sodium hypochlorite	Liquid chlorine from a drum
HCl	Hydrochloric acid	Acid most commonly used to lower pH
H_2SO_4	Sulphuric acid	Alternative acid used to lower pH
CO_2	Carbon dioxide	The gas that drives pH behaviour in pool water
H_2CO_3	Carbonic acid	A minor form of dissolved CO_2 in equilibrium with $CO_2(aq)$ typically present at ~650 times the concentration of H_2CO_3
HCO_3^-	Bicarbonate ion	The main alkalinity buffer in pool water
$NaHCO_3$	Sodium bicarbonate	Raises total alkalinity
H^+	Hydrogen ion	Determines acidity; directly affects ORP readings
OH^-	Hydroxide ion	Determines alkalinity; produced by salt chlorinators
H_2O	Water	—
CYA / $C_3H_3N_3O_3$	Cyanuric acid	Stabiliser that protects chlorine from UV; reduces active HOCl

This table is here for reference at any point.
There is no expectation that you will retain it after a single reading.

PART 1 — FOUNDATIONS

CHAPTER 5

What This Part Covers

This section sets up the why.

It covers the origins of this book, clears up common folklore, identifies where CO_2 works well and where it does not, and explains the so-called runaway TA, without blame or mystery.

If you are deciding whether CO_2 suits your site, start here. Come back to it later when something on the pool deck behaves unexpectedly.

The core mindset: pools are not static bodies of water. They are systems.

CHAPTER 6

Origins

The Start of Modern CO_2 Control

When I migrated to Australia in 1993, I already had a background in the swimming pool industry. But I knew enough to recognise that a new country meant new water conditions, new regulations, and new expectations. I contacted the state health department and asked who handled swimming pool chemistry. I was referred to Ty Webb, at the time the water scientist for the Government Chemical Laboratory.

Ty gave up his weekends to drive out to the outer northern suburbs of Perth to talk pool chemistry with me. In those sessions, he explained the role of carbon dioxide in pH control across all swimming pools and spas, particularly those with plastic equipment and fibreglass surfaces. Those conversations would later prove pivotal.

In late 1996, I was tasked with establishing the commercial division of our company, focused exclusively on commercial pool clients. I needed a point of difference in what was, at the time, a closed market. My enquiries turned up two aquatic centres already using CO_2. The gas supplier described escalating consumption at both sites, and Ty's explanations came back to me. I began thinking about how those principles applied to commercial aquatic centres of the materials, the plant room, the staff.

An idea formed: perhaps the industry was applying CO_2 incorrectly, and perhaps I could fix that in a way that staff would actually adopt. Armed with nothing but the idea, I picked up the phone.

The first call went to the centre closest to my home.

> *Hi, I'm David. I have an idea that could reduce your CO_2 consumption by at least 30%. Would that be of interest?"*

The reply:

> *"I wish you'd called last week. The CO_2 system was removed this morning — escalating TA and gas costs."*

It felt like being sunk before setting sail.
I made the next call anyway.

Same opening. Completely different response:

> *"What would you need from us to make that happen?"*

I explained that I would need the pool closed briefly to rebalance the water chemistry, and then approximately eight weeks of responsibility for calcium chloride and hydrochloric acid additions, working alongside their staff. Once complete, the team should be able to manage independently.

> *"How long would the closure be?*

> *"Two to three days."*

> *"We're closing for the Easter weekend. Will that be enough time, and what will it cost?"*

> *"Yes. I'd estimate $3,000 plus chemicals consumed."*

> *"Send me a quote."*

I spent that entire Easter weekend making the 35 km drive each way, often arriving at 5:00 am and leaving at 10:00 pm. I ran six to eight tests across the water bodies before making any dosing decision. When I was confident in the results, treatment began.

From memory, the initial addition was 1,000 litres of hydrochloric acid, appropriate for the volume involved. Total alkalinity at the start was 240 mg/L, and calcium hardness was around 100 mg/L. Before addressing calcium hardness, the total alkalinity had to come down first, to avoid precipitating calcium carbonate, a problem that cannot be reversed quickly without another large acid dose. I had until Easter Tuesday.

By Easter Monday, I was looking at a different pool. Water that had been grey and lifeless was now clear, bright, and inviting. When the centre reopened, bather numbers increased by around 70% within the week, not a gradual drift upward, but a noticeable jump.

I monitored tests and logbooks daily and phoned the liaison each day to walk through results and dosing decisions.

The immediate outcomes:

- CO_2 consumption dropped by more than two-thirds
- Aquatic centre revenue increased significantly
- An industry engineer who brought his daughter to squad training visited the centre, contacted me, and opened doors across the wider industry

Other aquatic centres followed. For the next two to three years, we installed bulk CO_2 dosing systems at a rate of two to three complete systems per month. By the time that market was largely satisfied, our state had around one million people, by one account, more CO_2 installations than any other state in the country.

I have continued refining my understanding since 1999. This book draws on that practical foundation from 1996–1999, updated through 2026. Early drafts written in 2023 were shared with a select group, including Richard Falk, who helped identify and correct scientific errors. His contributions are preserved in the appendices covering the Langelier Saturation Index and Henry's Law.

I hope that anyone who applies the principles here sees results as clear as those I witnessed over that Easter weekend in 1997, safer dosing environments, and safer water for the communities those centres serve.

CHAPTER 7

Reader Skill Level and Scope of This Guide

CO_2 pH-control essentials for pools and spas

This guide is written primarily for pool and spa professionals working with commercial aquatic systems, particularly those involved in carbon dioxide dosing for pH control. However, the chemistry, physics, and practical principles apply to any water body where pH control matters, from backyard pools to large public facilities.

Whether you are a technician, operator, maintenance professional, consultant, or an engaged homeowner seeking deeper understanding, this book offers practical guidance and clear explanations for working safely and effectively with CO_2 systems.

Some familiarity with basic water chemistry and pool equipment is assumed, though the step-by-step approach and real-world examples are designed to be accessible across a range of backgrounds.

Safe handling of CO_2 and pool chemicals requires proper training, compliance with local regulations, and respect for established safety protocols. This guide supports informed decision-making but does not replace formal certification or licensing requirements.

CHAPTER 8

Essential Safety Principles When Working with CO_2

CO_2 work: risk check, PPE, ventilation, no by passing safety

The procedures in this guide involve pressurised gases, pool chemicals, and associated equipment, all of which carry real risk. Carbon dioxide is non-flammable but is an asphyxiant. Improper storage, installation, or operation can result in serious injury or death.

This guide is intended for trained professionals who are familiar with local regulations and safe handling practices.

Always:

- Conduct a site-specific risk assessment before beginning any work
- Use appropriate PPE, including eye protection and gloves
- Follow manufacturer instructions and comply with all applicable laws and standards
- Ensure adequate ventilation wherever CO_2 is stored or used
- Never bypass safety devices or disregard warning signs
- Seek qualified advice if uncertain about any aspect of installation or operation

This guide supports sound decisions. It does not replace formal training or your site's regulatory obligations. When in doubt, ask someone with the appropriate authority.

CHAPTER 9

Folklore

Myths about CO_2 use and their explanations

Pick up this book or hear the phrase "carbon dioxide for pH correction", and a familiar set of concerns tends to surface:

CO_2 creates runaway alkalinity:

- It is expensive to manage
- It is difficult to install
- It makes the water cloudy
- It is only for large resorts and councils
- It cannot keep up with heavy bather loads
- It is slow to respond

There are likely more.

These concerns reflect real experiences. They are not invented. But they are driven by misapplication or misunderstanding, not by CO_2 itself.

This book works through each of them.

Folklore	Chapter
CO_2 creates runaway alkalinity	12
Expensive to manage	10,13
CO_2 is difficult to install	Part 2
CO_2 makes the water cloudy	Part 3
Is only for fancy resorts and councils	All
CO_2 can't keep up with heavy bathing use	All, and specifically 24,26
CO_2 is slow to respond	25

CHAPTER 10

When CO_2 Is Less Efficient

(Higher Consumption Likely)

Every aquatic facility already uses carbon dioxide for pH control — whether or not CO_2 gas is installed.

Consider this reaction:

$$HCl + NaHCO_3 \rightarrow NaCl + H_2O + CO_2$$

Hydrochloric acid + Sodium Bicarbonate
→ Salt + Water + Carbon Dioxide

When acid is added to water containing bicarbonate alkalinity, CO_2 is produced as part of the reaction. The pH reduction is occurring through the carbonic acid/CO_2 system, not directly from the acid. A facility cannot reasonably argue that CO_2 is unsuitable while continuing to dose acid for pH control. The underlying chemistry is the same.

CO_2 works in every pool. What varies is efficiency and gas consumption.

Facilities with high aeration or sustained turbulence will consume more CO_2 to hold the same pH setpoint. The following are indicators of higher consumption.

CO_2 Loss Spectrum
by Feature

Facilities With Continuous or Deliberate Aeration

Any feature that injects air into the water or breaks the surface will strip dissolved CO_2, increasing demand. Examples include:

- Spray pads
- Fountains
- Water cannons
- Tipping buckets
- Interactive play features

Water spray pads and highly active shallow play areas are generally poor candidates for efficient CO_2 pH control. The gas is released back to atmosphere almost as quickly as it is injected.

Spas

Heavy aeration from blowers and jets, combined with elevated water temperatures, accelerates CO_2 loss, increasing consumption and reducing control response.

CO_2 can still be used in spas. It is rarely, however, the most practical pH control method for heavily aerated installations.

Highly Turbulent Water Bodies
Sustained turbulence strips CO_2 continuously.

Examples include:

- Lazy rivers
- Rapid river attractions
- Wave pools
- Any pool designed to maintain constant surface disruption

Where injection, contact time, and control logic are not correctly configured, pH correction may lag behind actual demand.

Open-Air, High-Wind Facilities

Outdoor pools exposed to strong, consistent winds behave like low-level aeration systems. Rooftop pools, exposed coastal pools, and shallow leisure pools with extended beach entries are particularly affected. Even with all plant and equipment off, wind alone can drive meaningful water and CO_2 loss.

> *"One installation we completed sat on a building rooftop adjacent on the Swan River, open to the river on two sides. The Swan runs unobstructed on the ocean. Winds were severe enough that, with nothing operating, the pool water level would drop approximately 450mm overnight from evaporation alone."*

Toddler Pools

Toddler pools warrant specific consideration. They are typically:

- Very shallow, producing a high surface-to-volume ratio
- Warmer than main pool water
- Sometimes fitted with spray features
- Subject to constant surface agitation from bather activity

Higher gas consumption in these pools is not a system fault. It is a predictable consequence of their physical design and use.

CHAPTER 11

When CO_2 is More Efficient

(Apparently)

Some backyard pools using CO_2 for pH control operate without the runaway total alkalinity commonly reported in commercial settings. Owners often describe the system as working well with minimal intervention.

The chemistry is not different.

CO_2 behaves identically in a backyard pool and a commercial pool. What changes is the operating environment.

Compared to commercial water bodies, domestic pools typically have:

- Less turbulence
- Less aeration
- Lower bather loads
- Simpler automated dosing
- Less rigid TA targets
- Fewer or no water features

These factors reduce CO_2 loss and help stabilise pH.

When Domestic Pools Show the Same Problems

Some backyard pools do exhibit the same runaway total alkalinity seen commercially. These are typically pools that:

- Use continuous chlorine/pH controllers
- Are salt-chlorinated with TA above 60 mg/L[4.1]
- Active high aeration features (e.g. waterfalls, fountains, and spillovers) significantly increase CO_2 loss

In these cases, the operating environment begins to resemble a commercial pool, and the chemistry responds accordingly.

The Role of total Alkalinity

Where surfaces and equipment permit, domestic salt-chlorinated pools can often operate near 30 mg/L TA with notably improved pH stability and minimal acid demand.

This range is generally not available to commercial operators under current regulations. The domestic case nevertheless demonstrates the same principle: total alkalinity drives pH behaviour. The pH donor is secondary.

The Governing Principle

CO_2 is chemically identical in all pools.

Differences in performance reflect how quickly CO_2 is lost to the atmosphere, not any property of the gas itself.

CHAPTER 12

Runaway Total Alkalinity

CO_2 reveals — not causes — TA climb

This is one of the most misunderstood aspects of CO_2 pH control, and one of the most troubling for technicians and operators encountering it for the first time.

What's actually happening

In CO_2-dosed pools, total alkalinity climbs.

Left unmanaged, it can reach levels that look completely wrong — and feel alarming.

The obvious conclusion: CO_2 is causing the total alkalinity to rise.

That conclusion is understandable.

It is also incorrect.

Why it looks like a CO2 problem

CO_2 is the new variable. It gets installed, it gets switched on, and the TA starts climbing. The association is immediate and feels causal.

But the same chemistry is occurring in acid-dosed pools. The difference is that in those pools, it is not visible.

To understand why, two things need to be separated.

1. How pH is lowered
2. What happens to alkalinity in the background

What acid does

When a strong acid such as hydrochloric acid (HCl) is added to pool water, it releases hydrogen ions (H^+) directly. These immediately reduce pH, producing a fast and easily observed response.

In bicarbonate-buffered pool water, most of those H^+ are taken up by the carbonate system:

$$HCO_3^- + H^+ \rightarrow H_2CO_3 \rightarrow CO_2(aq)$$

Bicarbonate + Hydrogen → Carbonic Acid → CO_2 (aqueous)

The increase in dissolved CO_2 is a result of buffering, not the cause of the initial pH drop.

In summary, acid:

- Drops pH quickly
- Consumes alkalinity at the same time

What CO2 does

Carbon dioxide lowers pH by a different process. When CO_2 is injected into water, it dissolves and forms carbonic acid:

$$CO_2 + H_2O \rightleftharpoons H_2CO_3 \rightleftharpoons HCO_3^- + H^+$$

Carbon dioxide + water ⇌ carbonic acid ⇌ bicarbonate ion + hydrogen ion

This process forms a weak acid system in which hydrogen ions are released gradually and reversibly, producing a moderated pH response.

The end result looks the same — pH goes down. But the mechanisms differ.

Acid adds H^+ directly. CO_2 generates H^+ through the carbonate equilibrium already present in the water.

This difference also affects how far pH can be driven, as discussed in the pH control section.

This distinction becomes important when explaining why total alkalinity behaves differently under CO_2 control than under acid control

Why TA Behave Differently Under CO_2

Here's the part that causes the confusion.

In a real pool, chlorine (hypochlorite) is constantly being added. That produces hydroxide (OH-), which puts upward pressure on total alkalinity.

In acid controlled pools, hydroxide is neutralised by the added hydrogen ions, forming water.

In CO_2 controlled pools, hydroxide is converted into bicarbonate, which remains as part of total alkalinity.

In a CO2-controlled pool. CO2 does not neutralise OH- and the total alkalinity rise becomes visible.

The bottom line

CO2 is not causing the runaway TA.

It simply isn't hiding it.

The real driver is continuous hypochlorite dosing.CO_2 is not causing the runaway TA.

> *"Note:The worked examples in this book use 80 mg/L as the TA target because that is what regulations typically mandate. The appendices covering the Langelier Saturation Index and Henry's Law explain why lower targets are chemically superior. Where your regulatory framework permits it, that is the better path."*

What this means on site

- Expect periodic acid doses to manage TA in CO_2-dosed pools.
- CO_2 handles pH. Acid is still required to keep total alkalinity on target.

Why TA Appears to 'Run Away' with CO_2 Dosing

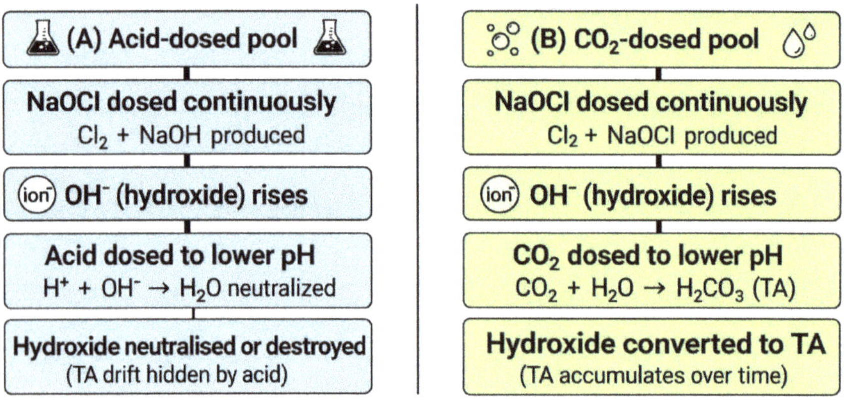

CO_2 does not hide alkalinity. Strong acid destroys hydroxide by converting it to water, while CO_2 converts hydroxide into bicarbonate.

CHAPTER 13

Expensive to Manage

Different expenses, similar totals

CO_2 is not an inexpensive product, and it requires appropriate storage vessels and detection equipment. However, overall costs are broadly comparable to those of maintaining an automated acid dosing system.

Acid systems often require dedicated ventilation, and their dosing equipment demands more frequent servicing and component replacement. In some regions, the cost of acid is not substantially different from that of CO_2 gas.

PART 2 — INSTALLING A CO_2 SYSTEM

CHAPTER 14

How to Use the Install Chapters

Follow chapters in order; test first, then build. Make small, logged adjustments only

Use these chapters as a build sheet.

Work through them in order: baseline testing first, then storage selection, tubing, sizing, protection, and labelling. Finish with diffuser placement and the commissioning checklist.

Do not make adjustments while you are still learning how the site behaves. Record everything you do. Small, deliberate changes produce better outcomes than large, clever ones.

CHAPTER 15

Before Starting Installation

Record full baseline chemistry first.

Before beginning any installation, establish the baseline chemistry of the water used to fill the pool.

Required tests:

- Free Chlorine
- Total Chlorine
- pH
- Calcium Hardness
- Total Alkalinity
- Cyanuric Acid (should be absent in the incoming water source)
- Total Dissolved Solids
- Any additional parameters specific to your location

Record all readings. If the local water supply is known to vary, consider collecting samples daily, weekly, or monthly over a period of time to capture that variation.

For example, in our region, the mains supply was historically a blend of groundwater and dam water, producing a variable incoming mix. In that situation, collecting weekly samples over four weeks and averaging the results gave a more reliable picture of what the water body could expect over time.

Once the incoming supply is documented, complete a full water test on the existing water bodies and record the same parameters.

These baseline values will be referenced throughout the pH control and operational chapters.

CHAPTER 16

Storage

Bulk CO_2: supplier-installed tank, ventilation, detection, protected regulators, correct manifolds, secure cylinders

CO_2 Storage Options at a Glance

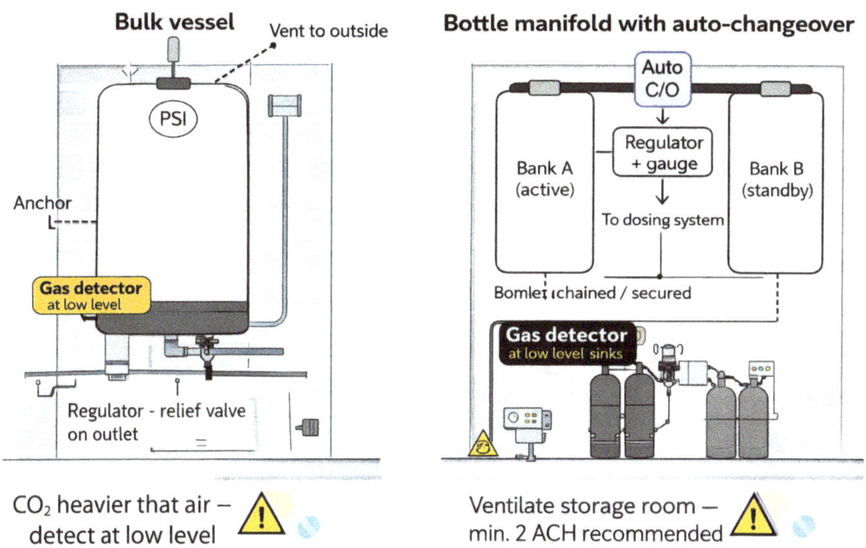

Bulk

In Australia, the process is straightforward: contact the gas supplier and arrange a site inspection to determine the best location for their vessel and associated equipment. The gas company typically installs the bulk vessel and runs a line to a regulator, from which the installer picks up the supply.

Arrangements in other regions may differ. The following principles, however, apply universally:

- Gas detection is mandatory in enclosed spaces
- CO_2 storage areas must be ventilated
- Regulators must be protected from moisture
- Changeover manifolds must be sized for the expected draw rate
- Bulk tanks require supplier involvement for installation
- Cylinders must be securely anchored and separated from ignition sources

These are true everywhere.

> *"Note: For all enclosed spaces, a sniffer or gas detection device is mandatory. This should be installed in the lower part of the room, as the gas is heavier than air. Always refer to the manufacturer's instructions and the local regulations for the specific requirements."*

Cylinder

Cylinder storage requires more planning. There is no reason it cannot be used effectively, but it is worth completing the consumption calculations in Chapter 29 first to confirm that cylinders are a viable option for the site's daily gas demand.

Location

In Australia, CO_2 is classified as a Class 2.2 non-flammable, non-toxic gas. As an asphyxiation hazard, the separation distances specified under Australian Standard AS 4332:2004 relate to:

- Preventing cylinder heating
- Preventing mechanical damage
- Preventing cross-hazards with other gas classifications:

Metre		Note
0	From ignition sources	CO2 is non-flammable no minimum distance required for ignition-source separation
≥ 1	From heat sources	Boilers, steam lines, hot surfaces, or radiant heat. Cylinders must be kept clear of heat sources that could raise cylinder temperature above safe limits.
3	From	CO2 (Class 2.2) must be separated from: • Class 2.1 flammable gases • Class 2.3 toxic gases • Class 5.1 oxidisers Note that the 3m can be measured around a 30-minute fire-rated wall
N/A	From property	• No fixed minimum distance from boundaries • But cylinders must be protected from vehicle impact and located where a leak cannot accumulate in pits, basements, or confined spaces.

As CO_2 is an asphyxiant, adequate ventilation is required.

The AS-mandated distances above are the minimum required on installation. The following additional separations are recommended:

Metric	Suggestion
3 metres	Minimum distance from door or window openings into plant rooms or other public spaces.
Avoid	Areas with pits or balance tanks below floor level.
Avoid	Adjacent to sub-surface workspaces, e.g. underground plant rooms.

CO_2 is heavier than air. It will accumulate in low-lying areas and confined spaces first.

Retain (Cylinder Security)

Gas cylinders are pressurised vessels. If a valve is knocked off through physical impact, the cylinder can become a lethal projectile. To prevent this:

- Place cylinders on a level, solid base
- Chain or strap them securely to a wall, bracket, or solid post
- Protect from impact on all sides using bollards, cages, or guard rails
- Protect from weather extremes, particularly direct heat and sunlight
- Never transport cylinders lying down unless fully restrained against movement

Ventilation

CO_2 is an asphyxiant. Ventilation deserves specific attention for this reason.

Where possible, storage should be located outdoors in a secure area with no direct path into a normal working space. Regardless of location, additional ventilation is advisable in any space adjacent to or through which gas lines are routed for pH control.

CO_2 may briefly appear as a cold fog on release. As it warms to ambient temperature, it becomes invisible and odourless.

Reminder: CO_2 sensors are typically mounted 150–300 mm above the finished floor level, near likely leak points. Always follow the detector manufacturer's instructions and the requirements of the authority having jurisdiction.

CHAPTER 17

Sensors and Alarms

Complex rooms = extra sensors

Gas detection is a life-safety control, not an optimisation tool. Its purpose is to alert occupants to dangerous accumulation before symptoms develop.

For outdoor installations, a gas detection alarm is not normally required under Australian Standards. For any installation inside a plant room or similar enclosed space, a gas detection system with alarms is mandatory.

In plant room locations, gas detection systems must be positioned:

- No more than 300 mm above the finished floor level
- To provide adequate monitoring across the entire plant room, including any pits, balance tanks, or below-floor areas

For pool plant rooms where storage is located outside but lines pass through the room to the diffuser, fitting an additional gas detection alarm inside the room is strongly recommended.

Room Dimensions / Characteristics	Typical Description	Suggested Number of CO_2 Sensors	Placement Notes
< 10 m² **(Small room)**	Small cellars, small cool rooms, small plant rooms	1 sensor	Mount low (≈ 600 mm above floor). Place as close as practical to cylinder(s) or leak source.
10–25 m² **(Medium room)**	Standard cellar,	1–2 sensors	Use 1 sensor if the room is clear and
	medium plant room		unobstructed; use 2 if shelving or partitions create dead zones.
25–50 m² **(Large room)**	Large cellars, larger indoor gas rooms	2 sensors	Place sensors on opposite sides of room, both low-mounted. Avoid placing both near doors/vents.
50–100 m² **(Very large room)**	Extended cellars, larger plant rooms	2–3 sensors	Space sensors evenly across room footprint. Prioritise corners or low-lying pockets where CO_2 can settle.
> 100 m² **(Complex / industrial room)**	Multizone plant rooms, long corridors, split areas	Minimum 3 sensors, often 4+	Treat each subarea or separated zone as needing its own detector. Add sensors near CO_2 piping

Room Dimensions / Characteristics	Typical Description	Suggested Number of CO_2 Sensors	Placement Notes
			manifolds or regulators.
Room with heavy shelving, pillars or machinery	"Broken up" rooms with poor airflow	+1 extra sensor beyond size-based number	Treat obstructed sections as separate zones. CO_2 settles and pools around obstacles.
Room with multiple CO_2 cylinders separated by distance	Two ends of a room, or two distinct storage points	1 sensor per cylinder zone	Each cylinder zone should have its own sensor due to local pooling patterns.
Room with gas lines entering from outside	Dosing lines, Beer gas systems, manifold lines	1 sensor near entry point + standard sensors	Detects leaks from piping, not only cylinder storage.
Any room with a cool room attached	Common in hospitality sites	1 sensor in cellar + 1 inside cool room	Treat coolrooms as separate atmospheres.

Pool plant rooms are not simple cylinder stores. Equipment, brackets, pipework, and other physical objects create obstructions that can interfere with normal sensor coverage. In complex plant rooms, additional sensors should be considered wherever CO_2 dosing lines are routed through or near other equipment.

CHAPTER 18

Auto Changeover and Regulator

Gotchas!

For bulk storage, the regulator is typically installed as part of the gas company's service.

For cylinder installations, the regulator is installed by the installer as part of the automatic changeover system.

An auto changeover head for CO_2 is a dual-inlet pressure regulator incorporating a small internal shuttle valve. The shuttle compares inlet pressures from two cylinders. Whichever cylinder carries the higher pressure automatically becomes the active supply. As the primary cylinder empties and all liquid CO_2 has boiled away, its pressure drops.

AutoChangeover

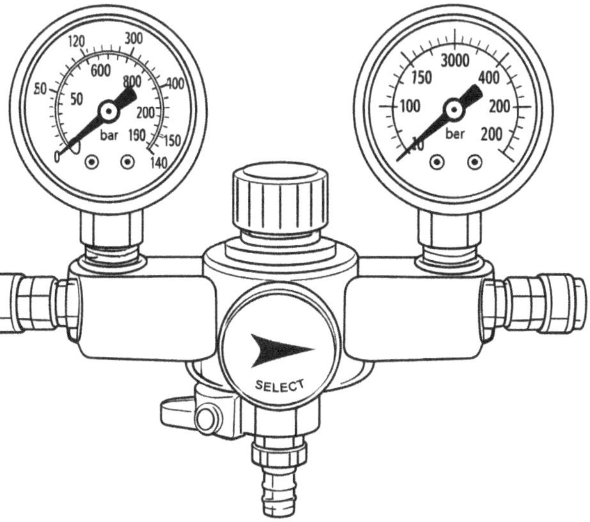

CO_2 is stored as a liquid under compression. As gas is drawn off to correct pH, the liquid boils, releasing gas for the dosing system. Because this process is governed by liquid–vapour equilibrium, the storage gauge will show near-constant pressure until the liquid is almost fully depleted, at which point pressure drops rapidly.

Always treat cylinders, whether apparently empty or full, as though they are full.

When inlet pressure falls below the regulator's differential threshold, the shuttle moves and the reserve cylinder takes over without interrupting the gas supply.

If you have seen how a Kreepy Krauly propels itself using pressure differential and an internal hammer mechanism, the shuttle valve in an auto changeover head works on the same principle. The side with greater pressure wins. When that pressure drops, the shuttle flips.

A colour window or arrow on the front of the module follows the shuttle's position mechanically, indicating which cylinder is active. The system requires no power, no sensors, and no calibration, only correct plumbing and properly tightened fittings.

CO_2 Regulator and Auto Changeover Gotchas

The following summarises practical issues that can catch technicians when working with CO_2 cylinders, regulators, and auto-changeover heads.

Gotchas Table

Gotcha	Explanation
CO2 is right-hand threaded	Australian Type 30 fittings use 0.860" WHIT, 14 TPI right-hand threads, not left-hand.
Washer required for Type 30	Flat-face sealing requires nylon, fibre, or metal-bonded washers between the regulator and valve.

Gotcha	Explanation
No PTFE tape	Tape must not be used on cylinder valve threads because sealing occurs at the washer, not the thread.
Don't mix standards	Type 30, CGA-320, DIN/BS CO2 threads are not interchangeable and can't be forced to fit.
Washer reuse limits	Fibre washers are single-use; nylon/metal washers may be reused if undamaged.
Over-tightening the nut	Too much torque damages the regulator seat or crushes the washer; firm compression is enough. Try hand-tight, then a quarter turn with a wrench. This is usually sufficient to create a positive seal on the washer/O-ring without crushing or distorting it.
Confirm correct stems & nuts	Ensure Type 30 stems/nuts are used, as Type 10/50 appear similar but won't seal correctly.
Do not mix BSP & NPT	Thread systems differ; mismatches can cause leaks or thread damage unless proper adaptors are used.
Autochangeover primary assignment	Set one side as primary and reset the indicator after replacing cylinders.
Pressure rating mix-ups	Use high-pressure-rated components upstream; many hobby fittings are only safe post-regulation.

Regulator

The regulator on a CO_2 storage system performs two key functions.

The first gauge indicates the pressure remaining in the active cylinder, whether that is a bulk vessel, a cylinder on an automatic changeover manifold, or a standalone cylinder. Because CO_2 is stored as a liquid under pressure, the storage gauge does not provide a linear measure of remaining volume. Pressure remains broadly constant while liquid is present, then drops sharply as the cylinder nears empty

Regulator

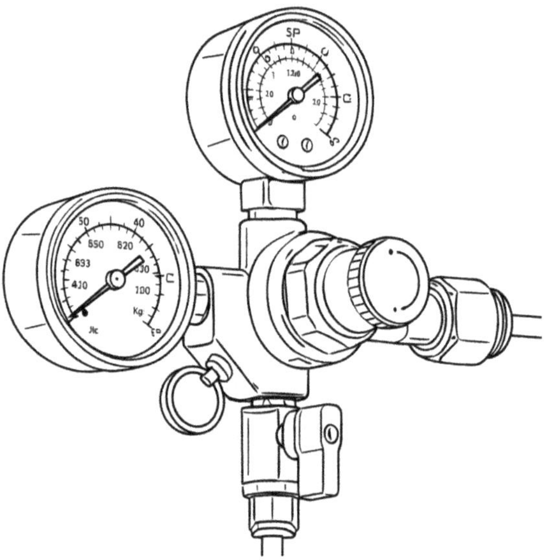

The second gauge displays the line or delivery pressure. This is typically adjustable via a control knob on the regulator body.

If delivery pressure is set too low relative to site demand, the system may draw gas continuously or at an excessive rate. Under these conditions, significant Joule–Thomson cooling can occur, observed as ice forming around the regulator body or around the solenoid valve in the CO_2 gas controller.

Some manufacturers offer heated regulators to address this.
In warmer climates, reducing the rate of gas draw is generally sufficient to eliminate icing.

CHAPTER 19

Tube and Fittings

Fittings Selection Basics

In the pool industry, the most common materials for chemical lines are PVC pipe and dosing tubes, typically PVC or cross-linked polyethylene, usually supplied as part of the chemical control package. Some CO_2 controller suppliers do not include tubes or fittings. This section provides the practical guidance needed in those cases.

CO_2 offers wider material compatibility than most pool chemicals. For cost-effective installations, nylon 12 tube is the standard choice. A local hose and fittings supplier will be able to confirm suitable options.

Push Fit Fittings

Tube fittings broadly fall into two types: push fit and knurl-and-nut.

Push fit fittings use a collet that is depressed to release the tube.

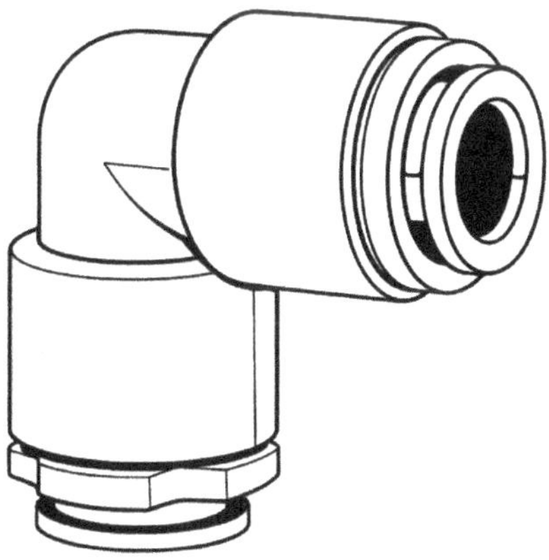

The tube is cut square and pushed into the fitting until it bottoms out. The O-ring provides the seal; the collet holds the tube in place.

Push fit fitting cut away

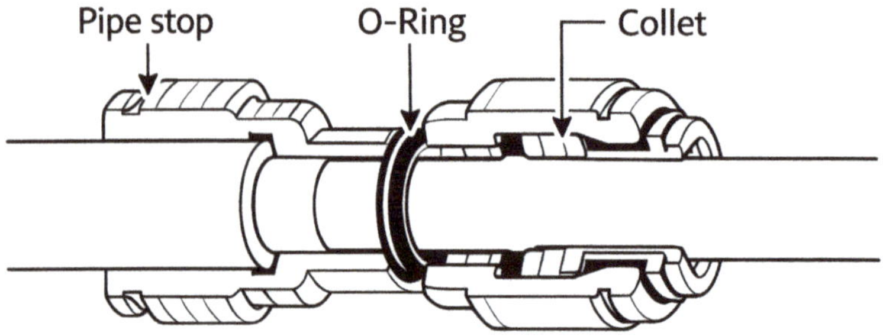

Push fit fittings are faster to install and reduce overall installation time. Before using, always confirm with the manufacturer that the fitting is suited to the specific installation, including chemical compatibility and the operating parameters of the pool system.

Knurl-and-Nut Fittings

The second type uses a knurl fitting, typically of metallic construction.

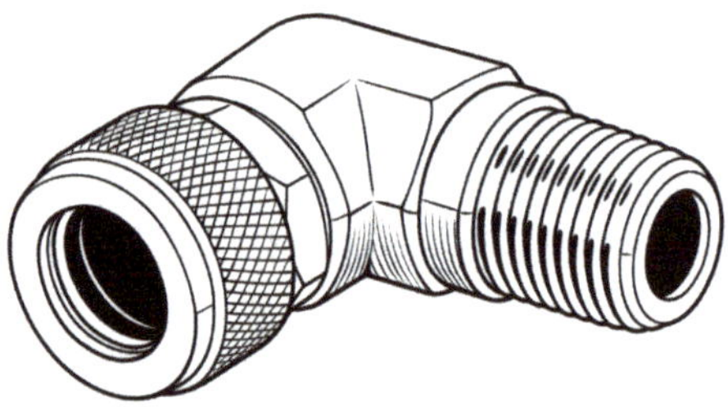

If metals are used, compatibility with all pool chemicals present must be confirmed. As these fittings are generally designed for industrial use, they are usually available in alloys suited to this application.

> *"Note: Always discuss the specific application with the tube and fitting supplier to confirm compatibility, not only with the chemical, but between the tube and fitting combination. Some fittings are designed exclusively for stainless steel tube; using nylon 12 tube with those fittings can result in the tube being severed. The supplier will advise which combinations are appropriate."*

These work differently from push fit. However, if in a hurry, they can be used similarly with experience.

Knurl Fitting Cutaway

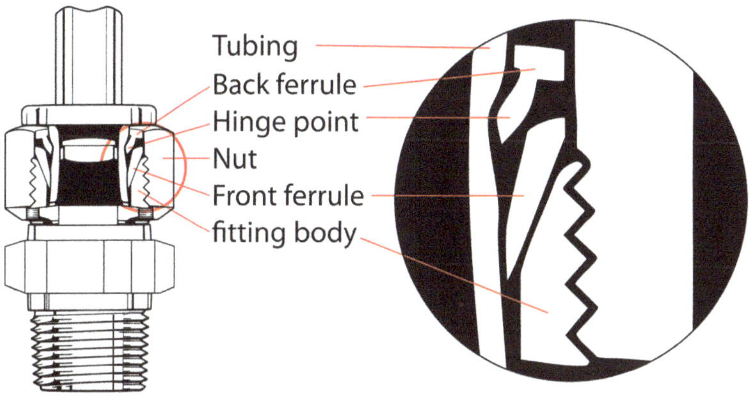

To install: loosen the nut, insert the tube until it bottoms out, then tighten the nut to seal. Always follow the manufacturer's instructions. This guide provides general direction only and is not a substitute for product-specific guidance.

Note: Tube used for CO_2 is not always compatible with other pool chemicals. Always verify suitability before use.

> *"Note: The tube used for CO2 is not always compatible with other chemicals. Always check materials for suitability before use."*

Sizing

Industrial tube and hose suppliers use different sizing conventions for metric and imperial tube. Metric sourcing is not guaranteed even in metric-primary markets.

Imperial sizing refers to outside diameter only ¼ inch tube has a ¼ inch external diameter. Metric sizing expresses both external and internal dimensions 8×5 means 8 mm outside diameter and 5 mm inside diameter.

For CO_2 dosing in pool applications, ¼-inch (or equivalent metric) nylon 12 tube is the most common choice.

Metric Tube Sizes and Nearest Imperial

Metric Tube (OD × ID)	Nearest Imperial	Notes
5 × 3 mm	~3/16" (4.76 mm)	Lightduty gas/dosing lines.
6 × 4 mm	~1/4" (6.35 mm)	Very common pneumatic/ dosing size.
8 × 5 mm	~5/16" (7.94 mm)	Common CO_2 feed line in the pool industry.
10 × 6.5 mm	~3/8" (9.53 mm)	General-purpose gas/liquid transfer.
12 × 8 mm	~1/2" (12.7 mm)	Used for long runs or higher flow.
15 × 11 mm	~5/8" (15.88 mm)	Less common but used in higher-capacity systems.

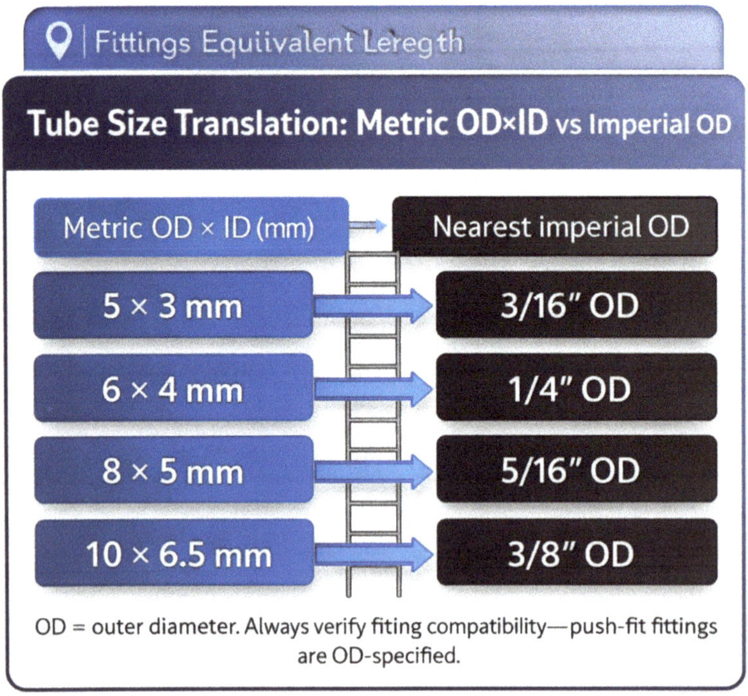

Calculating Flow

For detailed calculations, refer to the Appendix: Calculating Tube Size Example. The full working is laid out there. The mathematics has been made as clear and easy to follow as possible, no prior aptitude for numbers is assumed.

The Idea in one line

A longer or narrower tube slows gas flow. A larger inside diameter or a shorter run delivers more flow at the same regulator setting. Estimating Tube and Fittings Required

One practical way to estimate pipe length on site is to pace the distance. For most adults, a normal step covers approximately 150–300 mm; a long stride is close to one metre. Elevation changes can be estimated relative to your own height.

Alternatively, measure a repeating surface element, for example, a 450 mm paver across 14 units equals 6.3 metres.

Where accuracy matters, use a measuring tape, scale from drawings, or apply another reliable measurement method.

Before finalising the tube selection, walk the installation and record the likely fittings required downstream of the regulator:

Fitting	Count
90° standard elbow	
Swept long-radius bend	
Tee (straight through)	
Tee (branch take off)	
Ball Valve (full port)	

This covers the most common fittings used in CO_2 installations. If a specific fitting is not listed, estimate using the closest

Using the tables below and the fitting count recorded, the equivalent length can be estimated. Note that these values are generic. If the fitting supplier provides specific equivalent length data, use that in preference.

Fittings Equivalent Length Multiplier

Fitting Type	Averaged Multiplier (× D)	Notes
90° standard elbow	40 D	Midpoint of 30–50D range
Swept long-radius bend	18 D	Midpoint of 15–20D range
Tee (straight-through)	40 D	Middle of typical 20–60D range
Tee (branch take off)	80 D	Middle of 60–100D
Ball valve (full port)	30 D	Midpoint of 20–40D

Fitting Equivalent Length

Tube ID (mm)	40D Elbow (m)	18D Swept Bend (m)	40D Tee (straight) (m)	80D Tee (branch) (m)	30D Ball Valve (m)
4 mm	0.160	0.072	0.160	0.320	0.120
5 mm	0.200	0.090	0.200	0.400	0.150
6 mm	0.240	0.108	0.240	0.480	0.180
8 mm	0.320	0.144	0.320	0.640	0.240
10 mm	0.400	0.180	0.400	0.800	0.300

Before finalising the pressure drop calculation for fittings, refer to the Minimum Tube ID table below. Using the estimated total run length, select the appropriate tube size. In the worked example, the run is 27 metres above the 25 m column, so the 30 m column is used. The nominated regulator outlet pressure is 300 kPa.

The equivalent length method used here is conservative. Tube sizes are selected to keep CO_2 gas velocity below 15 m/s under expected operating pressures, providing margin for minor installation variations, additional fittings, and normal regulator adjustment. Where an installation falls outside these assumptions, verify using the full flow method in the Appendix.

With a 5 mm internal diameter tube confirmed from the Minimum Tube ID table, the worked example proceeds as follows:

Fitting	Count	Fitting Equiv. Length	Count x Fitting Equiv. L.
90° standard elbow	4	0.200	0.800
Swept long-radius bend	-		
Tee (straight through)	-		
Tee (branch take off)	-		
Ball Valve (full port)	1	0.120	0.120
Total			0.920

In this example, only a small number of fittings are present. In a more complex installation, this list will grow.

Add the fitting equivalent length 0.92 m to the measured tube run of 27 metres:

27 + 0.92 = 27.92 metres → round up to 30 m

This confirms a 5 mm tube at 300 kPa regulator pressure..

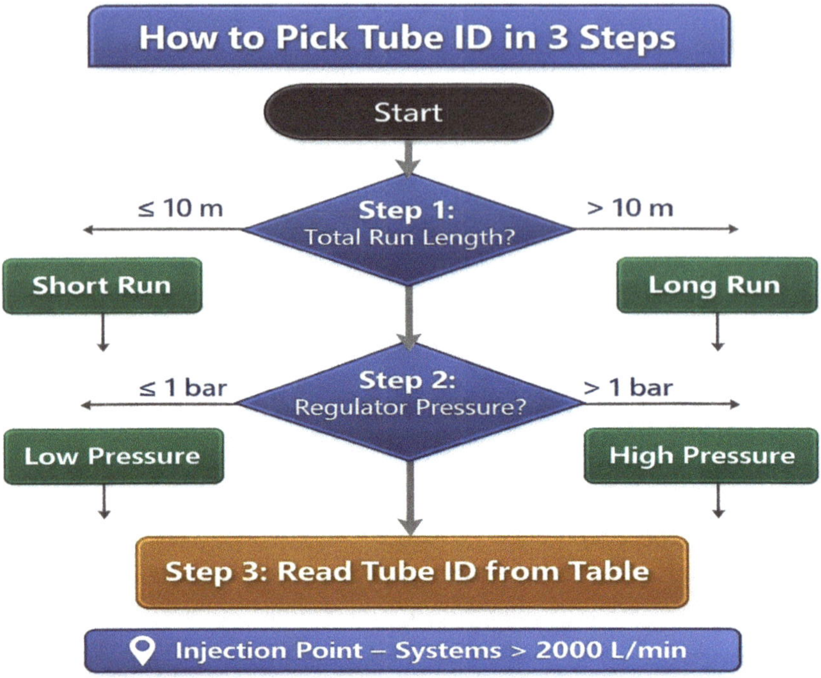

Proposed title: How to Pick Tube ID in 3 Steps

Minimum (plastic) Tube ID (mm) (using LP ceramic diffuser (See chapter 22 for diffuser types))

	Length m														
Reg P out (kPa)	5	10	15	20	25	30	35	40	45	50	60	70	80	90	100
100	6	6	6	6	8	8	8	8	8	8	8	8	8	8	10
150	6	6	6	6	6	6	6	6	8	8	8	8	8	8	8
200	5	5	5	5	5	6	6	6	6	6	6	8	8	8	8
300	5	5	5	5	5	5	5	5	5	5	6	6	6	6	6
400	4	4	4	4	4	5	5	5	5	5	5	5	5	6	6

This table shows the minimum internal diameter for smooth-bore polymer tubing, sized to keep CO_2 gas velocity below 15 m/s.

For metallic, rubber, corrugated, or any non-polymer tubing, internal roughness and friction factor will differ. In those cases, use the Appendix method to calculate the correct Reynolds number, friction factor, and resulting minimum tube ID.

Friendly rules of thumb

1. Bigger ID or shorter length → more flow with the same regulator setting.
2. If the response is slow, first try a larger tube or shorter run before changing controller tuning.

CHAPTER 20

Protection

CO_2 conduit: shielded, coloured, labelled

Encase the tube in conduit. Electrical conduit is a practical choice, it comes in convenient sizes for housing the tube, is widely available, easy to bracket, lightweight, and most importantly, protects the tube inside.

Standard 25 mm conduit is typically used for CO_2 lines.

At corners, flexible conduit replaces rigid bends. Flexible conduit accommodates changes in direction more easily than attempting to push tube through tight bends.

For conduit terminations, a cable gland adaptor provides a neat and secure finish.

Using conduit not only protects the tube but also serves a secondary safety function, if a leak occurs, the conduit partially contains the gas. However, electrical conduit is colour-coded for electrical use, not for gas service, so it requires painting before use as a CO_2 line.

Colour

In Australia, the identification of pipe, conduit, and duct contents is governed by AS 1345-1994, **Identification of the Contents of Pipes, Conduits and Ducts.** This standard specifies colours based on the content carried. Check local requirements for the applicable standard in your region.

Under AS 1345, the designated colour for CO_2 gas is Yellow Ochre / Sand. The specific colour reference under AS 2700:2011, **Colour Standards for General Purposes** is code Y44.

For an exact match, a dedicated paint supplier can produce a batch of AS 2700:2011-compliant Y44 paint on request.

There is, however, a practical concern with this colour: chlorine gas lines share the same Y44 designation. In an emergency, a CO_2 line painted to the standard colour could be mistaken for a chlorine line, a significant risk in a pool plant room where CO_2 is not always expected.

To address this, a sand tone is used in preference to the specified Y44, combined with clear labelling. This provides a visible distinction between CO_2 and chlorine lines within the plant..

Labelling

Always label the conduit with its content. A simple, purpose-printed label works well:

❮ **CARBON DIOXIDE** ❯

Trim the arrow from one end to indicate normal flow direction.

When ordering labels, specify vinyl outdoor permanent-adhesive material. This is waterproof and resistant to fading or damage in a pool plant room environment.

AS 1345 requires labels to be positioned so that the pipe content can be identified from a normal working position. Specifically, labels are required:

- At or near any valves
- At changes of direction
- At or near any junction
- At reasonable intervals on straight runs

For CO_2 conduit, labels are typically placed every 2.5–3.0 metres on straight runs. Where the conduit passes through a wall, bulkhead, doorway, or similar penetration, a label is placed on each side.

CHAPTER 21

Carbon Dioxide Gas Controller

Simple install, safe placement, solenoid freeze risk.

Various commercial pool dosing companies manufacture versions of this device. They are relatively straightforward, typically consisting of a solenoid valve, a gas flow rate meter, and an on/off switch.

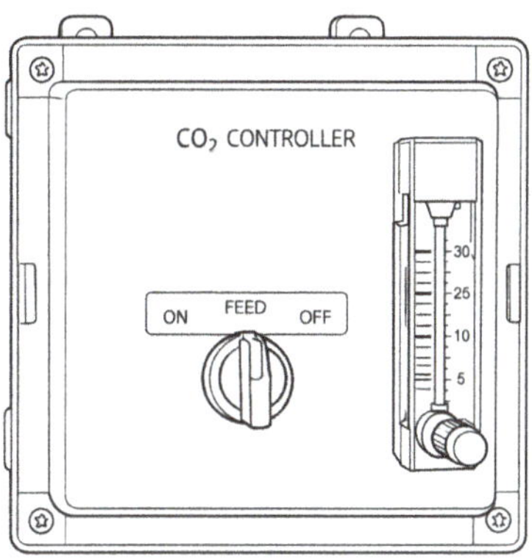

The internal workings are not complex and are covered later in the book.

For installation, a simple principle applies: the device includes a switch and an adjustment knob, and it displays a rate meter that will be read regularly. Install it at a height accessible to all staff, between knee and shoulder height of an average person. Given that the rate meter will be viewed frequently, err toward the upper end of this range.

> *"Note: Items that require only infrequent adjustment or are rarely read can be installed at a height suited to the task rather than to frequent access. For example, a flow-balancing valve set only at commissioning may reasonably require ladder access."*

Always install the CO_2 controller adjacent to the pH controller. CO_2 is a pH correction donor, so proximity is logical. As with a dosing pump, the pH controller can typically be used to engage the CO_2 controller manually for testing, being able to stand at the pH controller, switch to manual, and confirm the CO_2 controller is running is a practical operational advantage.

Like other electronic dosing devices, do not install the CO_2 controller where it is exposed to:

- Extreme heat or cold
- Excessive dust
- High humidity
- Overhead fluid risk
- Risk of impact from falling objects

The CO_2 controller requires electrical power. Depending on the pH controller model, power may be supplied directly from the pH controller, or a local general-purpose outlet may be needed with a signal cable running to the CO_2 controller. All electrical connections must be completed by a licensed electrician.

The tube connects to the CO_2 controller via either a purpose-specific fitting or, in some cases, a push-fit fitting integrated into the controller body. The outlet tube is then routed from the controller to the injection point.

If CO_2 demand is sustained or high, ice can form on the solenoid valve. This can freeze the solenoid in the open position, which increases gas draw, which in turn drives further Joule–Thomson cooling, a self-reinforcing cycle.

If pool pH is dropping sharply or holding around 6.1, a frozen solenoid is the most likely cause. Remove the CO_2 controller cover and inspect the solenoid. If it is coated in frost, apply a heat gun or similar device to melt the ice. Once warmed, the solenoid will close.

CHAPTER 22

Injection Point

“CO_2 diffuser fundamentals

The term “injection point” is in common use, but it is not particularly accurate, the device is almost always a diffuser or a venturi injector. Throughout this book, the term ***diffuser*** is used.

For CO_2 gas, the role of the diffuser differs from that of other chemical dosing points. This is where gas is introduced into the water. The goal is to introduce it as fine, evenly distributed bubbles not as a surge, as may occur with liquid chemicals such as chlorine or acid..

> *“Note: CO_2 is a chemical, even if it behaves differently from the liquids used in daily pool operations. Like any dosing point, the diffuser should be installed at least 350 mm from the chlorine injection point.”*

A ceramic diffuser is the preferred choice. Its relatively large surface area produces microscopic bubbles, maximising the interfacial contact area and promoting rapid, direct CO_2 absorption into the water..

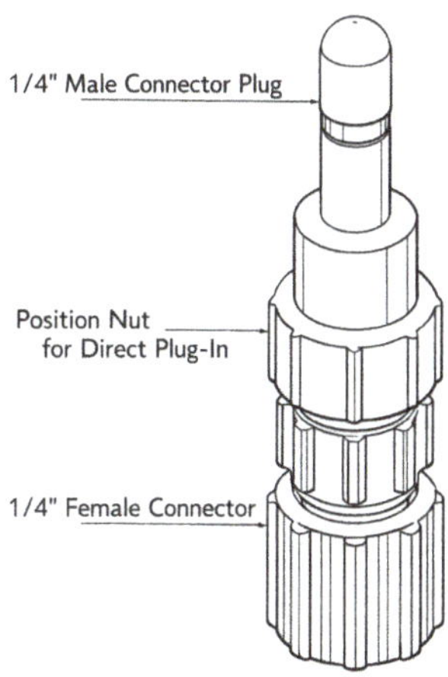

"Note: Fine bubbles visible at the pool return closest to the plant room are likely undissolved CO_2 that was not fully absorbed before reaching the return."

Placement

Diffuser placement is as important as diffuser selection. The following factors influence the decision:

- Bubble size: The guidance in this chapter assumes an LP ceramic diffuser. If a different diffuser type is used, bubble size must be factored into placement decisions.
- Gas hold-up: The mixing zone immediately surrounding the injection point affects how long gas remains in contact with water before the flow carries it away.
- Hydraulics:Turbulence and shear in the pipework assist dissolution.
- Temperature: Warmer water dissolves CO_2 more rapidly.
- Equilibrium point: This is the least understood factor and warrants explanation.

In this context, "equilibrium" means the point at which the concentration of dissolved CO_2 in the water is balanced by the rate at which CO_2 is leaving the water into the surrounding atmosphere. At equilibrium, dissolution and outgassing occur at the same rate.

Swimming pools almost always operate above the natural atmospheric equilibrium concentration of CO_2. Bather activity, heating, and turbulence all drive CO_2 out of solution faster than it can be replenished from the air. This has direct implications for pH behaviour, alkalinity, and aeration effects, and is discussed further in the operational chapters.

Ideal placement — high-flow systems (above 2,000 L/min)

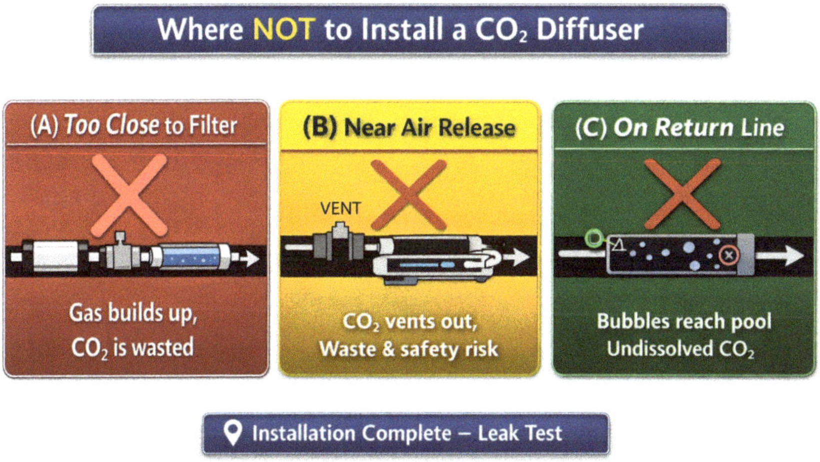

Systems with Flows over 2000 litres per minute

For high-flow systems, one location is markedly superior to all others: just before the pump inlet specifically, after the hair and lint pot and immediately before the pump impeller.

At this point, microbubbles are drawn into the impeller eye and subjected to intense mechanical agitation within the pump volute. This gives CO_2 the best possible opportunity for complete absorption, typically before the water exits the pump.

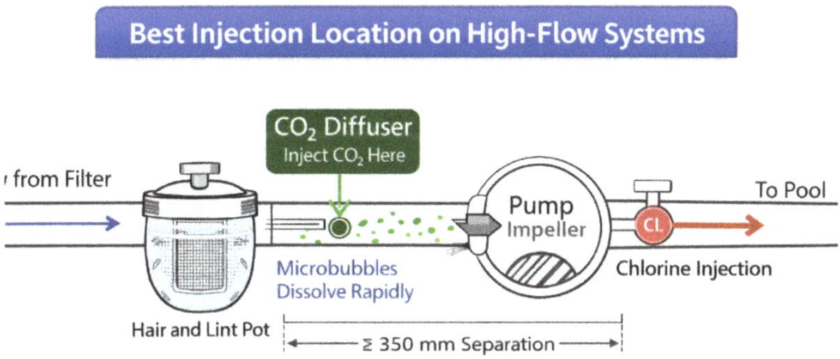

> **Caution**
>
> Maintain ≥ 350 mm separation between CO_2 and chlorine *injection points* to prevent reaction and gas evolution. Never inject CO_5 after chlorne in the same run.

The amount of CO_2 gas injected is usually very small. It is so small that at any time it won't create an issue for the pump.

Smaller Flow Systems

On smaller pumps, the hair and lint pot is typically moulded directly onto the pump body, making it impractical to install the diffuser between the pot and the impeller. An alternative location is required.

In these cases, pipe velocity becomes the key consideration. The relevant formula is:

$V = Q / A$

Where:

- V = velocity (m/s)
- Q = flow rate (m^3/s)
- A = pipe cross-sectional area (m^2)

Worked example:

Find V or velocity.

Flow rate from the pump curve: 500 L/min

Pipe diameter: 80 mm

Convert flow rate:

500 L/min = 8.33 L/s = 0.00833 m^3/s

Calculate pipe area:

Radius = 40 mm = 0.040 m

$A = \pi \times 0.040^2 = \pi \times 0.0016 = 0.00503\ m^2$

Calculate velocity:

V = 0.00833 ÷ 0.00503 = 1.65 m/s[14.1]

For quick reference in the field. Find the pipe diameter and the flow in litres per minute:

Velocity in metres per second by pipe and flow

L/min	40 mm	50 mm	80 mm	100 mm
300	0.994	0.636	0.248	0.159
350	1.160	0.742	0.290	0.185
400	1.326	0.848	0.331	0.212
450	1.492	0.954	0.373	0.238
500	1.657	1.061	0.414	0.265
550	1.823	1.167	0.455	0.291
600	1.989	1.273	0.497	0.318
700	2.321	1.485	0.580	0.371
800	2.652	1.697	0.663	0.424
900	2.984	1.909	0.746	0.477
1000		2.122	0.828	0.530
1200		2.546	0.994	0.636
1400		2.970	1.160	0.742
1600			1.326	0.848
1800			1.492	0.954

Lower pipe velocity indicates lower turbulence. CO_2 typically requires 5 to 35 seconds of contact time to be fully absorbed. Using the velocity table, the required straight pipe length for that absorption window can be estimated.

Worked example:

Pump flow: 550 L/min

Pipe size: 50 mm

Velocity from table: 1.167 m/s

At 5 seconds:

$5 \times 1.167 = 5.84$ metres

At 35 seconds:

$35 \times 1.167 = 40.85$ metres

In a small plant room, 40 metres of straight pipe is not practical. Options to address this include:

- Increase turbulence
- Reduce pipe diameter
- Increase pipe length where space permits
- Reduce bubble size
- Raise water temperature

In practice, a dedicated side-stream contact chamber is usually the most workable solution.

This bypass is not a standard system bypass, it is a dedicated loop designed specifically for CO_2 dissolution.

The configuration is as follows:

- A valve is installed on the discharge side of the pump
- A tee upstream of the valve diverts flow into a 25 mm diameter pipe
- A second tee downstream of the valve returns the flow to the main line
- CO_2 is injected just as the flow enters the 25 mm bypass from the larger bore pipe
- Approximately half the calculated absorption length is installed as a tightly bundled run of pipe and elbows, back and forth, within a compact footprint

For the worked example, approximately 18 metres of 25 mm pipe would be installed in this configuration.

TOP ELEVATION
SCALE 1:10

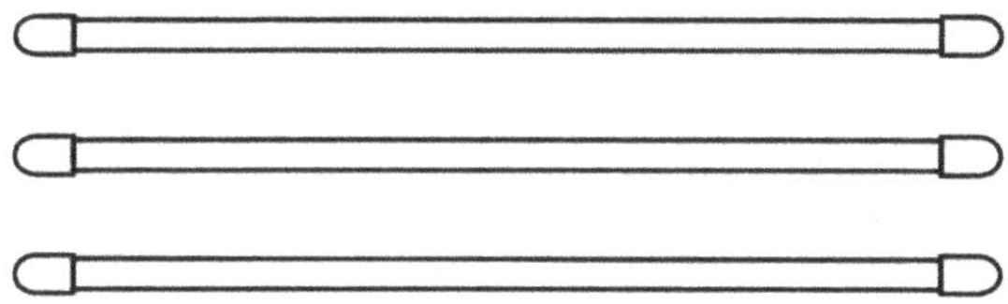

PLAN
SCALE 1:10

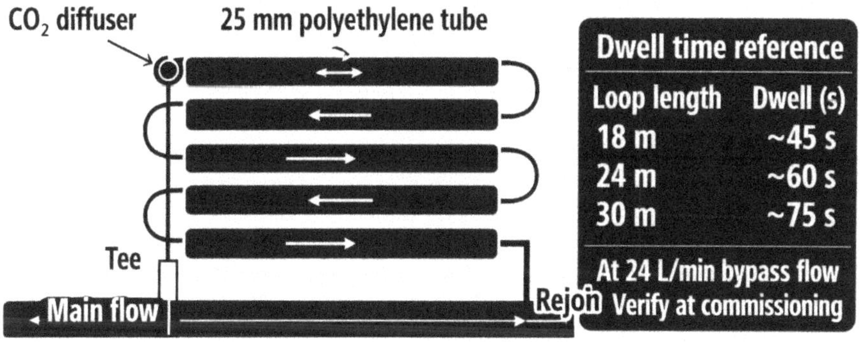

Main pipe (50–90 mm typical)

Bypass proportion: typkally 5–10% of main flow

More passes = longer contact time = better absorption.

The contact chamber accelerates flow velocity in the smaller diameter pipe, increases turbulence, and creates a tortuous path that extends contact time, all of which assist CO_2 absorption. The principle is straightforward. Making it look neat takes practice.

Pipe Length	*Volume (L)*	*30 L/ min*	*45 L/ min*	*60 L/ min*	*75 L/ min*	*88 L/ min*
18 m	8.8 L	17.6 s	11.7 s	8.8 s	7.0 s	6.0 s
24 m	11.8 L	23.6 s	15.7 s	11.8 s	9.4 s	8.0 s
30 m	14.7 L	29.4 s	19.6 s	14.7 s	11.8 s	10.0 s

To set the bypass flow rate: partially close the valve between the bypass legs until a pressure increase of approximately 10–15 kPa is read downstream of the pump.

⚠ **Caution:** Do not install CO_2 injection points immediately upstream of filters, air release valves, or late in the return circuit. Undissolved gas can accumulate and vent at pool inlets. On high-flow systems, prefer the pump eye location. Where that is not practical, use a dedicated side-stream contact chamber.

Where not to place the diffuser

Immediately upstream of the filter is a poor location. If CO_2 is not fully absorbed before the water slows inside the filter, gas will collect at the filter crown and vent through the air release mechanism.

Similarly, positioning the diffuser immediately upstream of any pipework air release valve will result in gas venting directly rather than dissolving.

Avoid placing the diffuser late in the return circuit. If installed on the return side after all other equipment, undissolved CO_2 is likely to bubble up at the pool return inlets.

CHAPTER 23

Installation Complete

CO_2 system verification

The preceding chapters have covered the components and installation sequence for a complete CO_2 dosing system.

At this point, the physical installation should be complete. Before proceeding to start-up, the system must be tested.

Pressure test:

Open the gas supply without starting any other equipment. The new lines should pressurise, and the reading should stabilise.

Walk the entire installation, listening for any hissing sounds. Hissing indicates a potential gas leak and is a quick, low-effort first check.

Where fittings are accessible, either by pulling back conduit sleeving or at exposed joints, apply a small amount of soapy water to each connection. Bubbles forming at a joint confirm a gas leak at that point.

Only after the installation has been tested and checked should the CO_2 controller be activated. Depending on the manufacturer, there will be a manual activation method use it, and repeat the soapy water test on all joints downstream of the controller.

Once the soapy water test is complete and any issues have been resolved, the full system is ready for start-up.

Configure the chemical controller for the facility, enable automatic CO_2 control, and set the CO_2 flow meter. As a starting point, set the ball at two-thirds of scale, or use the manufacturer's recommended setting.

> *"Note: The gas storage and supply connection requires a licensed gasfitter and compliance documentation in most Australian states. Your gas supplier will typically manage this. The broader dosing installation tubing, controller, and diffuser has no specific national licensing or sign-off requirement. Responsibility sits with the facility operator and their WHS framework. Always confirm current requirements with your state gas regulator and local health authority. As a matter of good practice, notify the relevant regulator by email when a change has been made to the facility's chemical dosing system."*

PART 3 — OPERATING A CO_2 POOL

CHAPTER 24

Day-to-Day Operation in Plain English

Tune gently, trim TA, cut aeration, trust experience

Tune gently. Hold pH within a band, trim total alkalinity with small, deliberate acid doses, and reduce CO_2 demand where possible by minimising aeration and wind exposure.

The CO_2 consumption calculator is a guide, not a guarantee. Trust the operator's lived experience, the best data often walks around in steel-capped boots.

What Drives CO_2 Loss

The rate of CO_2 outgassing increases under the following conditions:

- Higher total alkalinity
- Lower pH
- Increased aeration or turbulence

For lowest operational cost, CO_2 systems perform best at:

- Lower total alkalinity
- Higher pH within allowable limits
- Minimal unnecessary aeration

pH Control

Observe, drift, adjust

CO_2 pH control behaves differently from traditional acid dosing. Acid produces an immediate pH change because hydrogen ions are added directly to the water.

What this means for the day-to-day operation
CO_2 lowers pH through dissolution and equilibrium. The gas must first dissolve, form carbonic acid, and then release hydrogen ions. This makes the response slower and more gradual than acid dosing.

Watch the pool, not just the screen.

Science behind this (Why the above works)

The "softer" response of CO_2 compared to acid comes from how it behaves as a weak acid.

A weak acid does not release its hydrogen ions freely. Its effect on pH is limited by its dissociation constant (pKa), which for carbonic acid is around 6.3. By comparison, hydrochloric acid has a pKa of about −1 and releases hydrogen ions almost completely.

This matters in practice. As pH falls toward this point, carbonic acid becomes less effective at releasing additional hydrogen ions. The pH response slows.

A strong acid behaves differently. Hydrochloric acid continues to release hydrogen ions regardless of pH, steadily driving pH lower and consuming total alkalinity.

CO_2 does not reduce total alkalinity, so the buffering capacity remains in place. That limits how far the pH can fall, even under overdosing conditions.

For this reason, a CO_2 installation should not be considered complete once the hardware is commissioned. The chemical control still needs to be matched to the specific behaviour of the pool, and this is best done in close consultation with the operator who manages the controller day to day.

> *"Note: The regular operator will know more about how their pool responds than anyone else. Wherever possible, defer to their judgement on the adjustments that follow."*

Three rules for controller adjustment

Chemical control adjustments should follow three simple rules:

First rule: Do not adjust the chemical controller.

Second rule: If adjustment is required, make changes incrementally and slowly.

Third rule: Adjust one thing at a time, and record both the original setting and the new setting before making any change.

The first rule may appear contradictory if changes are required, the settings must eventually be adjusted. The intent is not to prevent adjustment, but to delay it until sufficient observation has revealed what actually needs to change.

Before touching the controller

The guidance below assumes a pool that has been operating normally and that an acid dosing pump is being replaced with CO_2. A brand-new installation where the chemical controller has not yet been commissioned will require a different approach.

Before reviewing any controller settings, observe the pool itself. Note whether the following are active:

- Fountains
- Weirs
- Spray points
- Any other features aerating the water

Take note of bather load, pool depth, and how the pool is currently being operated. Speak with the operators, is the pool, as you see it, in a typical state, unusually quiet, or busier than normal? These observations provide essential context before any settings are reviewed.

Next, observe the pH reading on the controller while watching the pool. Where practical, monitor both simultaneously. The goal at this stage is not to intervene but to understand how changes in bather load and pool activity are reflected in the pH reading.

Pay attention to:

- How quickly pH responds
- Whether the reading is stable or erratic
- How the controller calls for correction,frequent on/off cycling, or slow and sporadic response

After observing for at least five minutes and preferably longer, review the controller settings.

Control philosophy

Many controllers are configured with a rigid on/off response. For example, if the setpoint is 7.5, the controller doses whenever pH rises above that value.

A more flexible approach generally produces better results. Rather than enforcing a tight response at a single value, set a target pH and allow controlled drift around it. Subject to regulatory approval, there is usually no reason pH cannot drift between, for example, 7.35 and 7.65. These limits can typically be configured within the controller's proportional, integral, and derivative settings.

pH Band Control vs On/Off Tight Setpoint

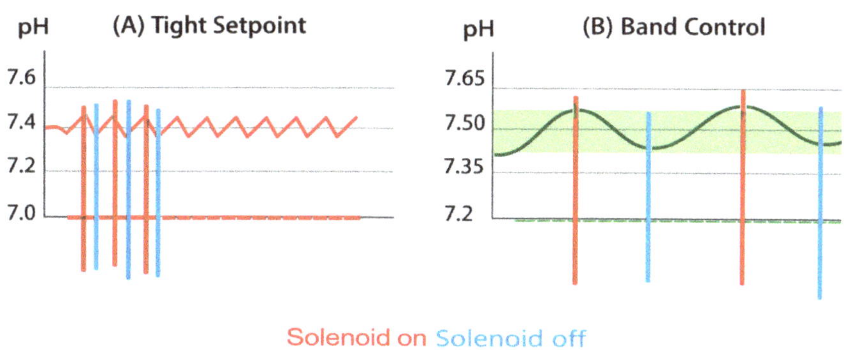

Both graphs show the same average **pH**: Band control reduces actuation frequency and COs micto-dosing losses.

Record all pH controller settings before making any changes.

> *"Note: This book does not document controller-specific menus or parameters. For detailed guidance on individual controllers, refer to the manufacturer's documentation."*

Once the pH controller has been configured based on observation and the manufacturer's instructions, enable the CO_2 system and allow it to begin controlling pH. Observe operation for at least an hour, watching how pH responds and making only minor adjustments if required.

Once the pH controller has stabilised, record the following:

- Free Chlorine
- Total Chlorine
- pH
- Calcium Hardness
- Total Alkalinity
- Cyanuric Acid
- Total Dissolved Solids

These values can be added to the baseline results recorded before installation, creating a comparative dataset for ongoing reference.

Leave the site and return the following day. An early morning visit works well. Compare the controller settings against those recorded at start-up, and complete another water test using the same parameters.

Pay particular attention to any change in total alkalinity. A rise greater than approximately 20 mg/L indicates a meaningfully different operating condition compared to a marginal increase. This will vary between pools, chlorine consumption is directly responsible for alkalinity rise, and pools with higher bather loads consume more chlorine, generating more hydroxide and increasing total alkalinity more rapidly.

If the facility is willing, ask them to include total alkalinity in their daily log. A once-per-day TA reading will be useful in the chapter that follows.

CHAPTER 26

Correcting Total Alkalinity

Chlorine drives alkalinity.

You test a pool three days in a row and see the same result each time: total alkalinity is climbing, despite nothing being added to deliberately raise it. When you mention it, a senior technician shrugs and says, “That’s just CO_2 pools.”

That explanation is common. It is also wrong.

CO_2 does not raise total alkalinity. The rise is real, but the cause lies elsewhere, and it can be corrected once it is properly understood.

This chapter explains how to bring total alkalinity back under control in CO_2-managed pools. Doing so requires accepting one key principle that underpins everything that follows.

As established in Chapter 12, the alkalinity rise is caused by continuous hypochlorite dosing not by CO_2.

This chapter includes calculations. They are unavoidable when adjusting alkalinity correctly, but they are not intended to be intimidating. The worked examples are provided so that, when needed, you can open the book, substitute your own numbers, and work through them step by step.

Science behind this

Continuous hypochlorite dosing introduces more than chlorine to the water. Each dose also adds a small amount of alkaline material.

Corrected: Chlorine and Alkalinity

When chlorine is consumed, it does not generally increase total alkalinity.

Most chlorine reactions either:

- Have **no net effect on alkalinity,** or
- **Reduce alkalinity slightly**

Examples include:
Photolysis:

$$2OCl^- \rightarrow O_2 + 2Cl^-$$

Two hypochlorite ions → oxygen + 2 chloride ions

Ammonia oxidation:

$$3OCl^- + 2NH_3 \rightarrow N_2 + 3Cl^- + 3H_2O$$

Three hypochlorite ions + two ammonia → nitrogen gas + three chloride ions + three water

Urea oxidation:

$$3OCl^- + \text{urea} \rightarrow N_2 + CO_2 + \text{products}$$

Three hypochlorite ions + urea →
nitrogen gas + carbon dioxide + products

Key point:

The rise in total alkalinity comes primarily from:

- **Excess sodium hydroxide (lye) in chlorinating liquid**
- **Fill water during evaporation and refill**

This is why total alkalinity rises steadily in pools with continuous hypochlorite dosing.

CO_2 does not remove this alkalinity. It adjusts pH by shifting the carbonate balance, it has no lasting neutralising effect on the base being continuously introduced by chlorine dosing.

In acid-dosed pools, this alkalinity rise is often hidden. Acid is added for pH control, and in doing so, it simultaneously neutralises the accumulating alkalinity. The rise is masked.

In a CO_2-controlled pool, that masking effect is absent:

- CO_2 controls pH only
- It does not neutralise alkalinity
- The alkalinity rise therefore becomes visible over time

This is why operators often conclude that CO_2 is causing the TA to rise. It is not. CO_2 is simply allowing what was already happening to become visible.

Because of this, acid still has a role in CO_2 pools, but that role is now deliberate:

Acid is used to control total alkalinity, not pH.

Taking a Total Alkalinity Reading

To calculate the correct acid dose, begin with a total alkalinity reading.

> *"Note: Water bodies are imperfect circulators, not all areas of the pool will read the same TA. When testing a water body for the first time, take six or more readings from different locations and depths and average the results rather than relying on a single reading. Once the pool's behaviour is better understood, a single reading may be sufficient."*

The following are also required:

- Pool volume
- Acid percentage strength (available on the SDS, the label, or both)

If the label states a range, for example, 31–37% use the lower figure for the first calculation.

Understanding "Total Alkalinity as Equivalent Calcium Carbonate"

> **Important Note on Units**
> *All dosing factors in this section are based on* **weight (grams)**.
>
> *If dosing by volume* **(millilitres or litres):**
>
> - *The density (specific gravity) of the acid must be used for conversion*
> - *Values in litres will differ from values in grams*

Although commonly referred to simply as Total Alkalinity, the full technical term is Total Alkalinity expressed as equivalent calcium carbonate.

This phrasing causes confusion. It is worth understanding clearly.

A helpful analogy: consider the phrase "the sea is as blue as the sky." This does not mean the sea is the sky, or the sky is the sea. It means their colours are comparable on the same scale.

Alkalinity works the same way.

Expressing alkalinity as calcium carbonate does not mean the alkalinity in the water is caused by calcium carbonate. It means the combined alkalinity of the water is expressed as if it were equivalent to a certain amount of $CaCO_3$. This provides a single, standardised unit for measurement.

In a laboratory, the individual contributors to alkalinity primarily bicarbonate, carbonate, and hydroxide, and in some cases borate or cyanurate anions, can be measured separately. In field water treatment, these are measured together and reported as one combined value: Total Alkalinity (as $CaCO_3$).

Using calcium carbonate equivalence allows different alkaline species, each with its own chemical strength to be reported on a single, standardised scale. This is why total alkalinity can be calculated, compared, and corrected using straightforward equations.

This concept matters beyond the immediate calculation. As operators progress in their careers, they will encounter laboratory reports that list individual anions rather than a combined total alkalinity value. In those cases, the alkalinity contribution of each species must be converted to equivalent calcium carbonate before the data can be correctly interpreted.

Understanding total alkalinity as equivalence, rather than as a substance is the first step toward confidently reading laboratory reports, reconciling field and laboratory data, and making informed chemical control decisions.

Calculate Acid Dose

To calculate how much acid is needed, use a fixed multiplier called a factor.

This factor represents the amount of acid required to reduce total alkalinity by 1 mg/L in 1,000 litres of water.

For the acids most commonly used in pools:

- 32% hydrochloric acid (HCl): **2.28**
- 28% sulphuric acid (H_2SO_4): **3.50**

These numbers do not need to be calculated on site. They are fixed multipliers, simply use them.

Practical method (use this on site)

1. Determine how much TA needs to drop
2. Multiply by the factor
3. Multiply by pool volume in thousands of litres

Worked example

Pool volume: 1,300,000 litres Acid: 32% HCl → Factor: 2.28 TA reading: 97 mg/L TA target: 80 mg/L

Step 1 — TA reduction required 97 - 80 = 17 mg/L

Step 2 — Apply the factor 17 × 2.28 = 38.76 → round up to 39

Step 3 — Scale to pool volume 39 × 1,300 = 50,700 mL = 50.7 litres

The factor already accounts for acid strength, chemical behaviour, and unit conversions. There is no need to convert to kilograms or apply specific gravity when using this method.

Optional fully worked explanation

The following section shows how the factor is derived and how it applies to both acid types. This is background for those who want to understand the mathematics behind the multiplier.

The equivalent weight (ew) is a standard chemistry term describing the mass of a substance, in this case, an acid in grams, that supplies one standard unit of reactive hydrogen ions (H^+). It allows different acids to be compared on a common basis.

Equivalent weights for the acids used in pool water treatment:

Acid	Equivalent Weight (ew)
HCl Hydrochloric or Muriatic	36.46
H_2SO_4 Sulphuric or Non-Fume	49.04

If another acid is used, its equivalent weight can be looked up and substituted into the same calculations. The method does not change, only the equivalent weight.

Worked example – starting values

What we know:

- Pool volume: 1,300,000 L
- TA reading: 97 mg/L
- TA target: 80 mg/L
- Acid: 32% HCl

> *"Note: If cyanuric acid is present in the pool, correct the TA reading before calculating. Subtract CYA ÷ 3 from the TA reading. For example: TA 100, CYA 50 → corrected TA = 100 – (50 ÷ 3) = 83.3 mg/L. Cyanuric acid is not used in this example."*

Full Calculation (Optional Background)

The practical method gives you the answer you need on site. This section shows the full working behind it, where the factor comes from and how the final volume is derived. It is optional reading.

Example A: 32% Hydrochloric Acid (HCl)

What we know:

- Pool volume: 1,300,000 L
- TA reading: 97 mg/L
- TA target: 80 mg/L
- Acid: 32% HCl

Daily/Three-Daily TA Correction – Workflow

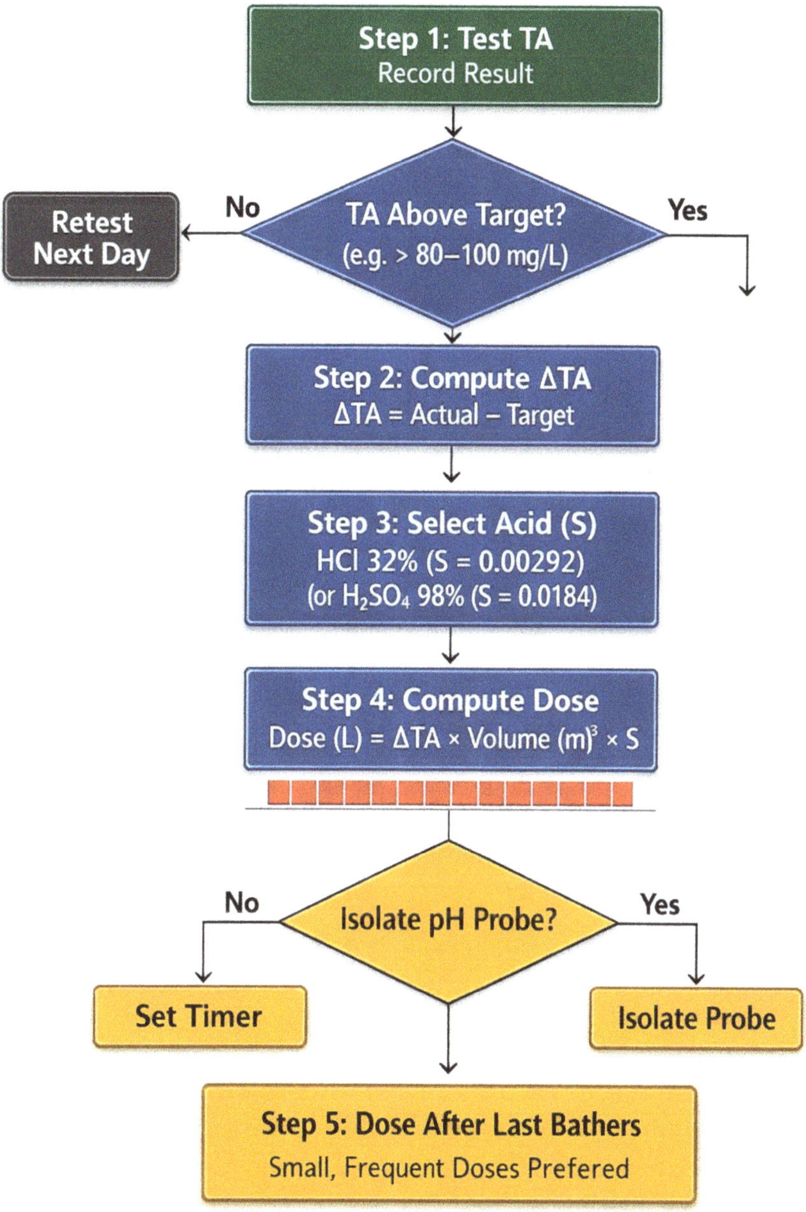

Always add acid to water, never water to acid.

Wear PPE. Never dose while pool is occupied.

Step 1: TA correction required

Total Alkalinity reading – Total Alkalinity target =
Total Alkalinity correction

97 mg/L – 80 mg/L = 17 mg/L

Step 2: Derive the factor

The factor is calculated once and reused every time this acid strength is applied. This step shows where it comes from.

Two values are needed:

- **EW** (equivalent weight of HCl) = 36.46
- **S** (acid strength as a decimal) = 0.32

Total alkalinity expressed as $CaCO_3$ is based on chemical equivalence. By convention, 50.045 g of $CaCO_3$ represents one equivalent. This is the standard reference unit that allows different alkaline species to be compared and corrected on a common basis.

The factor is calculated as:

$$F = EW \div (50.045 \times S)$$

$$F = 36.46 \div (50.045 \times 0.32)$$

$$F = 36.46 \div 16.014$$

F ≈ 2.28 mg acid per mg TA

This value is rounded upward by approximately 0.3%. While early rounding is not standard scientific practice, the difference is negligible at pool scale and is appropriate for field use.

For 32% hydrochloric acid, the factor is always 2.28. It does not need to be recalculated, only reused.

Also from the SDS, look up the specific gravity of the acid under physical or chemical properties:

9. PHYSICAL AND CHEMICAL PROPERTIES	
Appearance	Form: Liquid Colour: Clear, colourless to slightly yellow
Odour	Pungent
Odour Threshold	No data available
pH	<1
Melting Point	< −20 °C
Boiling Point/Range	98 °C (for 28 % concentration)
Decomposition Temperature	Not available
Evaporation Rate	No data available
Flash Point	Not applicable
Flammability Limits	Not applicable
Specific Gravity	1.14 (for 28 % concentration) 1.16 (for 32 % concentration)
Vapour Density (air = 1)	1.3
Vapour Pressure	2 kPa
% Volatiles	100
Solubility in Water	Miscible with water

ρ = **1.16 kg/L** for 32% HCl

Step 3: Dose per litre

Multiply the TA correction by the factor:

17 mg/L × 2.28 = **38.8 mg/L**

This is the mass of acid required per litre of pool water to achieve the correction.

Step 4: Total mass

Scale to the full pool volume:

Mass (kg) = (38.8 × 1,300,000) ÷ 1,000,000

Mass = **50.44 kg**

9. PHYSICAL AND CHEMICAL PROPERTIES	
Appearance	Clear colourless
Odour	Slight odour
pH	< 1 @ 10% solution
Boiling Point	Not available
Melting Point	3 °C
Evaporation Rate	Not available
Flash Point	Not applicable
Flammability Limits	Not applicable
Specific Gravity	1.24 – 1.27
Vapour Density	Not available
Vapour Pressure	Not available
Solubility (water)	Soluble
Upper Explosion Limit	Not applicable
Lower Explosion Limit	Not applicable
Partition Coefficient	Not available
Auto Ignition Temperature	Not available
Decomposition Temperature	Not available
Explosive Properties	Not available
% Volatiles	Not available

Step 5: Convert mass to volume

Using the specific gravity from the SDS:

Volume (L) = Mass ÷ ρ

Volume = 50.44 ÷ 1.16

Volume ≈ 43.5 litres

Relative to the total pool volume of 1,300,000 litres, this is a small and controlled dose.

Example B — Same scenario but H_2SO_4

Sulphuric acid, sold in Western Australia as non-fume acid, is available in a range of strengths.

> *"Note: If the acid label states a range such as 31–37%, always use the lower figure. This results in a slightly larger dose than strictly necessary, which is safer than underdosing."*

What we know:

- Pool volume: 1,300,000 L
- TA reading: 97 mg/L
- TA target: 80 mg/L
- Acid: 28% H_2SO_4

Step 1: TA correction required

97 mg/L – 80 mg/L = **17 mg/L**

Step 2: Derive the factor

Two values are needed:

- **EW** (equivalent weight of H_2SO_4) = 49.04
- **S** (acid strength as a decimal) = 0.28

$$F = EW \div (50.045 \times S)$$

$$F = 49.04 \div (50.045 \times 0.28)$$

$$F = 49.04 \div 14.013$$

F = 3.50 mg acid per mg TA

As with hydrochloric acid, this calculation is performed once and reused. For 28% sulphuric acid, the factor is always 3.50.

Step 3: Dose per litre

17 mg/L × 3.50 = **59.5 mg/L**

Step 4: Total mass

Scale to the full pool volume:

59.5 × 1,300 = **77,350 mg per litre total → 77.35 kg**

Step 5: Convert mass to volume

Sulphuric acid is supplied as a liquid. Using a specific gravity of approximately 1.25 kg/L at 20 °C:

Volume = 77.35 ÷ 1.25

Volume ≈ 61.9 litres → rounded to 62 litres

The acid dose required to correct 17 mg/L of total alkalinity in this pool is approximately **62 litres,** a small and controlled quantity relative to the total pool volume.

CHAPTER 27

How to Dose Acids

Safe, Controlled Dosing

"Note: Acid must only be added after all bathers have left the water. Never dose during operating hours."

TA calculated acid dose can be applied by several methods:

- Pouring directly into the gutter or balance tank
- Dosing via a chemical dosing pump
- Using a venturi system acid is measured into a container and drawn into the pool
- Other site-specific methods

Of these, a dosing pump is generally preferred. It delivers acid in a controlled manner, improves operator safety, and reduces the risks associated with manually handling larger volumes.

Because dosing pumps and timers are not infallible, it is good practice to physically limit the maximum volume of acid that can be delivered in any single dosing cycle. This provides a secondary safety layer in the event of equipment malfunction or incorrect programming.

The following procedure outlines a practical, conservative approach:

1. After the last bathers have left the pool, set up the required volume of acid using appropriate safety procedures.
2. Set a dosing pump timer based on the calculated dose. For example, using the HCl calculation above, 43.5 litres at a pump rate of 15 litres per hour, the required dosing time is approximately 3 hours.

3. Set the drum or container to the calculated volume plus a small margin to help maintain pump prime. In the worked example, setting up 60 litres means the maximum acid that could be delivered is 60 litres. Given a TA target of 80 mg/L, this provides a safety buffer against overnight overshoot if the dosing system runs longer than intended.

4. Start the dosing pump and confirm by observation that it is operating correctly.

"Note: Depending on how the injection system is configured, it may be necessary to isolate the chemical probes overnight. However, because the acid volume is small relative to pool volume and is being added slowly, significant pH spiking is unlikely. This can only be assessed by a trained professional familiar with the site's dosing control systems."

CHAPTER 28

Minimising Consumption

CO_2 use: cut disturbance

Anything that splashes, blows air, or adds wind increases CO_2 loss. Reducing aeration and wind exposure, allowing pH to sit slightly higher within regulatory limits, and operating at a lower total alkalinity all reduce gas demand.

One of the most common criticisms of CO_2 pH control is high gas consumption. In practice, this is largely a system design and operation issue, not a flaw in CO_2 itself.

Why this works:

Any increase in surface area or shear, fountains, blowers, wind accelerates CO_2 loss through gas exchange. Less outgassing means less CO_2 is required to hold the same pH band.

CO_2 Consumption Reduction Matrix

Action	Impact	Cost	Notes
Pool cover overnight	HIGH ★	Low	Stops overnight off-gassing
Raise pH setpoint band	HIGH ★	Nil	7.4–7.8 (typical regulatory range). Where precise control is possible, operating toward the upper end of This range (e.g. 7.6–7.8) reduces CO_2 demad.
	HIGH ★	Nil	
Lower TA target (80 mg/L)	HIGH ★	Nil	Less buffer to overcome. Check local requirements and aim for lowest possible level.
Reduce fountain / aeration runtime	MED	Nil	Balance with attraction needs.
Upgrade to membrane diffuser	MED	Med	Higher dissolution efficiency.
Wind break / roof glazing	MED	High	Capital works — site specific.

★ = high impact / low cost — prioritise these first.

Reduce aeration

CO_2 leaves pool water through gas exchange at the surface, a process known as gaseous interchange. This is a normal part of how water behaves.

Anything that increases aeration accelerates this loss:

- Fountains
- Blowers
- Splashing
- Sprays
- Tipping buckets
- Water slides

In practical terms, increased aeration increases the effective surface area of the water. More surface area means more CO_2 escapes to atmosphere, which in turn increases the gas required to maintain the same pH band.

In theory, the most CO_2-efficient water body would resemble a still lake minimal surface disturbance, simple skimmer collection rather than wet-deck grates. In practice, this is neither realistic nor desirable.

Pools are active, welcoming spaces. Reducing CO_2 consumption is therefore a matter of compromise, not elimination balancing gas efficiency against the function and enjoyment of the water body. What that compromise looks like will vary by site, patron type, and operating schedule. The goal is not to remove features, but to operate them intelligently.

For example, sprays and tipping buckets in a toddler pool might be placed on a timer, operating only during peak use periods, perhaps 9:30 am to noon on weekdays. Outside those hours, they can remain off without affecting the bather experience.

Where patron use is irregular rather than time-based, a push-button activation is effective. A single press could enable features for a fixed interval, say 10–15 minutes, before automatic shut-off.

Similarly, water slides may provide little benefit during quiet weekday periods and may only need to operate during weekends, school holidays, or after-school hours.

There are often many small, site-specific adjustments like these that, taken together, can significantly reduce CO_2 demand without compromising how the pool is used.

Use of Cover

Gaseous interchange occurs at the water surface. Using a cover when the pool is not in use will significantly improve gas retention and reduce demand.

However, on nights when acid is being dosed to correct total alkalinity, covers should be left off, outgassing on those nights is desirable..

Reduce wind

In this context, wind simply means air moving across the water surface. This affects indoor pools just as much as outdoor ones.

Where possible, minimise air movement directly across the surface. This recommendation applies outside of chloramine removal requirements, air handling for chloramine reduction must not be compromised.

Beyond that requirement, air should be redirected away from the pool surface. Options include:

- Windbreaks for outdoor pools
- Alternative diffuser positioning or deflection strategies for indoor facilities

> *"Note: For indoor facilities, a qualified HVAC professional familiar with chloramine management must be consulted before making any adjustments to existing air-handling systems."*

Temperature Consideration

Higher temperatures increase both CO_2 solubility and CO_2 loss.

In practice, reducing temperature is rarely an effective control strategy, as changes must be large enough for bathers to notice.

More practical methods include:

- Reducing aeration
- Lowering total alkalinity
- Operating at a higher pH target

Raise pH

Water with a lower pH contains more dissolved CO_2. Raising pH reduces dissolved CO_2 concentration, and with less CO_2 in solution, less is available to outgas, and less needs to be replaced.

If the regulated pH range is 7.2–7.8, allowing pH to sit closer to 7.7 is acceptable. Always operate within regulatory limits, as these are regularly checked.

Lower TA target

Total alkalinity can be reduced significantly below common industry norms. A TA of 30 mg/L will require substantially less CO_2 than a TA of 80 mg/L. This approach is chemically sound but is not always permitted by local regulation. Any departure from mandated targets must be documented and formally approved. For further detail, refer to the appendices on LSI and Henry's Law.

Reduce chlorine levels

As the primary source of hydroxide in pool water, anything that reduces chlorine demand will reduce CO_2 demand.

If the regulated free chlorine range is 3.0–8.0 mg/L, consider a target of 3.5 mg/L rather than 4.0 mg/L. Alternatively, consider allowing chlorine to vary within a band, for example, 3.5 to 4.5 mg/L rather than holding a fixed target. A more relaxed range may assist both regulatory compliance and CO_2 reduction.

Reduce Surface Disturbance

Bather activity will always disturb the water surface. Any realistic reduction in surface disturbance will improve CO_2 efficiency, even marginal improvements contribute over time.

Plant Maintenance

Not all CO_2 loss is driven by visible water features. Poor plant condition can also increase gas consumption, often without being immediately obvious.

Any condition that introduces air into the system or allows dissolved gas to escape before it can equilibrate will increase CO_2 loss. Examples include:

- Pump cavitation, which draws air into the water stream
- Air leaks on the suction side of the system
- Leaking or poorly seated air release valves
- Changes to pipework or equipment that allow gas to escape before it dissolves

Cavitation and air ingress are typically discussed in terms of pump performance and mechanical wear. They are also, however, a direct cause of gaseous interchange and increased CO_2 loss a connection that is easy to miss.

Plant maintenance is not just a mechanical concern. It is often directly connected to chemical consumption.

CHAPTER 29

CO2 Consumption Calculator

CO_2 use: calculated guide

This section provides a practical method for estimating likely CO_2 consumption for a pool. The calculation begins with chlorine demand and accounts for real-world losses including aeration, turbulence, and incomplete gas dissolution.

The result is a guide, not a guarantee. Actual consumption measured on site will always take precedence. No calculation can fully capture the specific behaviour of an individual water body.

The preceding chapters have covered installation, start-up, total alkalinity control, and consumption reduction. What has not yet been addressed is what normal or reasonable CO_2 use actually looks like. Is a kilogram per day too much? Too little? The best way to answer that is to work through a realistic example.

Assumptions

These are the assumptions being used for this example:

- Pool volume: 1,300,000 litres
- Chlorine: maintain 4.0 mg/L target, with a loss of 1.0 mg/L per hour (This assumes a steady feed to hold 4.0 mg/L. Real pools are not this uniform, but the simplification is useful for demonstration.)
- Chlorine donor: Sodium Hypochlorite 12.5%
- pH band: 7.4 to 7.7

- Total Alkalinity: 80 mg/L as $CaCO_3$
- CO_2 injection efficiency: 50% (only that fraction dissolves and reacts)
- Average aeration/turbulence treated as typical and already reflected in the injection efficiency figure

How Sodium Hypochlorite Increases pH

As established earlier, hypochlorite introduces hydroxide into the water, which drives both pH rise and CO_2 demand.

$$NaOCl + H_2O \rightarrow HOCl + NaOH$$

Sodium Hypochlorite + Water →
Hypochlorous Acid + Sodium Hydroxide

The sodium hydroxide produced increases both pH and total alkalinity.

As a practical rule of thumb:

- Each 1 mg/L of free chlorine added increases alkalinity by approximately 0.6–0.8 mg/L (as $CaCO_3$)
- Neutralising 1 mg/L of alkalinity (as $CaCO_3$) requires approximately 0.88 mg/L of CO_2

Using the midpoint of the alkalinity range, the stoichiometric CO_2 requirement per unit of chlorine is:

0.6–0.8 mg/L alkalinity × 0.88 mg/L CO_2 per mg/L alkalinity ≈ 0.62 mg/L CO_2 per mg/L chlorine

This value represents a chemical reference point only, based on ideal neutralisation and no system losses. It does not determine actual CO_2 demand in real pools.

In practice, CO_2 demand is driven by outgassing, which is influenced by aeration, turbulence, surface disturbance, temperature, total alkalinity, and pH.

In practice, CO_2 dosing requirements are best determined by measuring actual consumption over time and adjusting for operating conditions.

"Note: This conversion applies specifically to sodium hypochlorite. For calcium hypochlorite, use 0.71. For other chlorine sources, consult the supplier's chemistry data."

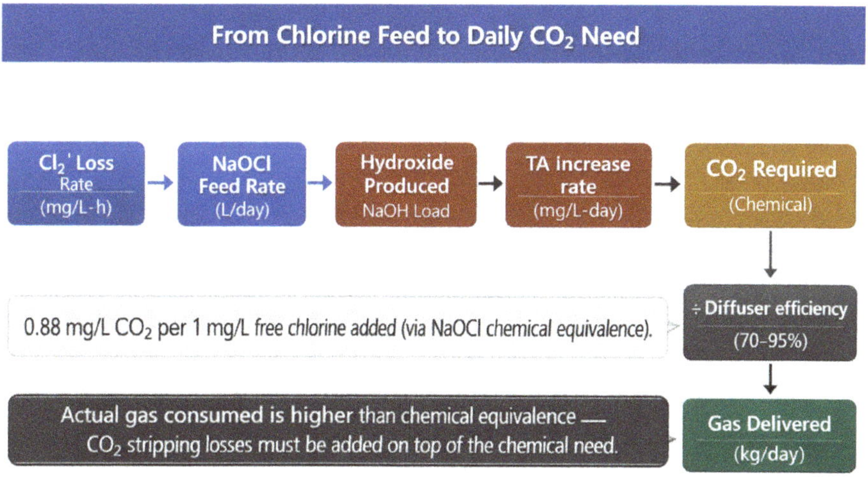

Daily Chlorine Loss → Daily CO_2 Requirement:

1.0 Chlorine loss rate: 1.0 mg/L per hour × 24 hours = **24.0 mg/L per day**

"Note: To measure actual chlorine consumption on a site using bulk sodium hypochlorite storage:"

1. Mark the sight glass at the current liquid level. Record the date and time.
2. The following day at the same time, mark the new level. Record the date and time.
3. Repeat for several days or weeks.
4. Measure the distance between marks and use the tank diameter to calculate daily volume consumed.
5. Average the daily figures for a reliable consumption estimate.

CO_2 required at the water to offset the hypochlorite effect:

24.0 mg/L chlorine × 0.62 mg/L CO_2 per mg/L chlorine = 14.88 mg/L CO_2 per day

Convert mg/L to kg/day

.Pool volume: 1,300,000 litres

Total CO_2 mass per day at 100% dissolution: 14.88 mg/L × 1,300,000 litres = 19,344,000 mg/day

Convert to kilograms: 19,344,000 ÷ 1,000,000 = 19.344 kg/day

At perfect dissolution and zero losses, the pool requires 19.344 kg of CO_2 per day.

Account for CO_2 Injection Efficiency

Not all CO_2 fed into a pool dissolves and reacts. Some is lost through incomplete dissolution, bypassing, or surface blow-off. This is described as injection efficiency.

For this example, an efficiency of 50% is assumed, typical for an LP ceramic diffuser installation. This means only half the supplied CO_2 is assumed to dissolve and contribute to pH control. The remainder is effectively lost.

This assumption is deliberately conservative. Actual efficiencies may be higher, but a pessimistic value avoids underestimating real-world consumption.

Chemical CO_2 requirement ÷ injection efficiency = delivered gas per day

19.344 kg ÷ 0.50 = 38.688 kg/day

Daily CO_2 gas feed estimate (at 50% efficiency): approximately 38.7 kg/day

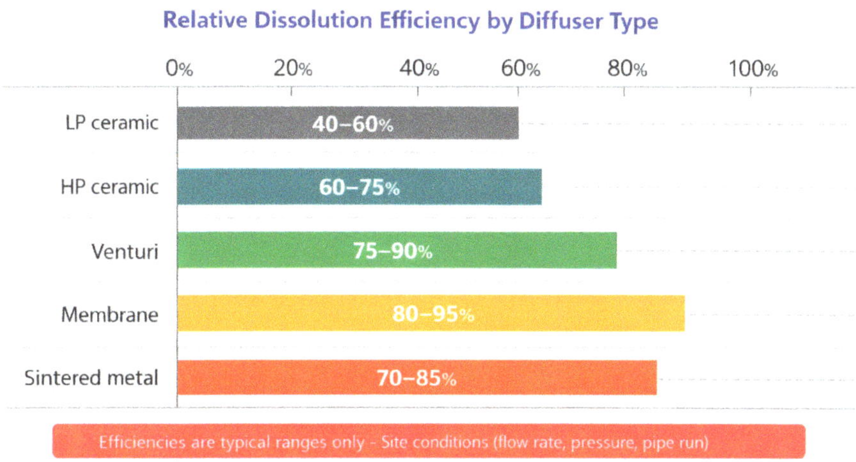

Diffuser Type	Typical CO_2 Dissolution Efficiency	Notes (Real-World Behaviour)
LP Ceramic (Low-pressure ceramic disc/	35% – 55%	Produces moderately fine bubbles. Very dependent on placement; performs best in low-turbulence side stream contactors.
HP Ceramic (High-pressure micron	60% – 85%	Excellent gas transfer. True "fine pore" ceramics can exceed 80% if submerged at depth with low turbulence. Ideal for CO2.
Venturi Injector	20% – 35%	Good mixing, but coarse bubble size and minimal contact time in most pool plumbing.
Membrane Diffuser (EPDM or PTFE)	40% – 70%	Bubble size varies with pressure and membrane age. New membranes perform closer to 70%. Good compromise between performance and maintenance.
Sintered Stainless Diffuser	50% – 75%	Produces very fine and consistent bubbles. Performs best when operated at stable pressure above ~2 bar. Durable and stable efficiency.

Summary

Accurately predicting CO_2 consumption is difficult because some variables cannot be captured in a simple model. Fill water chemistry, backwashing practice, controller algorithms, temperature, and bather load all influence gas demand. Any one of these, or a combination can shift actual consumption away from the calculated value.

For this reason, it is generally better to overestimate likely consumption than to chase precision. This calculation is a guide, not a guarantee.

CHAPTER 30

CO_2 pool field troubleshooting

CO_2 pool troubleshooting quick guide

These checks assume the system has been correctly installed and that all safety systems are functional. Always prioritise site safety and regulatory requirements. Work through checks in order. Change one thing at a time. Record what you changed.

pH keeps rising — CO_2 not holding it down

Check 1: First action; Open the CO_2 controller cover. Is the solenoid covered in frost or ice? If yes, use a heat gun to melt it. A frozen solenoid may be stuck open or stuck closed. pH plummeting to around 6.1 indicates frozen open. pH rising despite the system running indicates frozen closed.

Check 2: Is the system calling for CO_2? Watch the pH controller for at least 5 minutes. Is it triggering the solenoid? If not, review the controller setpoint and band settings.

Check 3: What is the total alkalinity? High TA requires more CO_2 to hold pH. If TA is above target, dose acid first. CO_2 controls pH; acid controls TA.

Check 4: Is aeration running? Fountains, sprays, tipping buckets, and blowers all strip CO_2 from the water. Higher CO_2 demand when features are active is normal, not a fault.

Check 5: Are bubbles visible at pool returns? Undissolved CO_2 is escaping before it can act. Review diffuser placement, it should not be immediately upstream of filters or air release valves.

Normal behaviour: CO_2 response is softer and slower than acid. If pH is within regulatory limits and moving predictably, observe before intervening. Small drift is not failure.

pH crashing too low (sitting ~6.1 or below)

Check 1: First action; Open the CO_2 controller cover. Frost on the solenoid? Yes, the solenoid is frozen open and feeding continuously. Use a heat gun immediately to close it. This is the most likely cause.

Do not attempt to correct low pH in a CO_2-controlled pool using sodium carbonate (pH increaser).

Sodium carbonate raises both pH and total alkalinity. In CO_2-controlled systems, this introduces additional alkalinity that the system will not remove, increasing future CO_2 demand and destabilising control.

First identify and correct the cause (commonly a frozen solenoid or overfeed). Once gas feed is stopped, pH will typically recover naturally as CO_2 outgasses.

Check 2: Is the CO_2 flow rate set too high? Reduce the flow meter setting. CO_2 response is slow, allow time between adjustments before assessing the result.

Check 3: Is the controller band too tight, for example, on/off at exactly 7.5 rather than allowing drift? Widen the band to, say, 7.35–7.65 if regulations permit.

Total Alkalinity keeps climbing week after week

Note, this is normal: TA rising in a CO_2 pool is not caused by CO_2. It is caused by continuous hypochlorite dosing adding hydroxide. CO_2 makes it visible; acid would have masked it.

Action: Dose acid to return TA to target. Acid controls TA; CO_2 controls pH. Small regular doses, daily or three times per week, are more effective than one large monthly addition.

Action: Calculate the acid dose: (TA reading − TA target) × pool volume, using factor 2.28 for 32% HCl. Dose after the last bather. Set a timer, do not free-pour.

If persistent: Consider reducing the chlorine target slightly within the regulatory range, or reducing aeration features during off-peak hours. Both reduce the hydroxide load.

Gas consumption is very high

Check 1: Is aeration running more than necessary? Fountains, sprays, blowers, and water features continuously strip CO_2. Timer or push-button controls on features can reduce demand significantly.

Check 2: Is the pool covered when not in use? A cover dramatically reduces overnight CO_2 loss. Note: leave covers off on nights when dosing acid for TA correction.

Check 3: Is total alkalinity running high? High TA increases CO_2 demand. Bring TA to target with acid, consumption should fall.

Check 4: Are there air leaks in the plant room plumbing? Cavitation, leaking air release valves, or recent pipework changes can all indirectly increase gas demand.

Check 5: Is the regulator delivery pressure set correctly? Pressure set too low for site demand causes continuous rapid gas draw, leading to Joule–Thomson icing on the regulator.

No gas reaching the pool / system not dosing

Check 1: First action; Is the cylinder or bulk supply empty? Check the regulator gauge, pressure drops rapidly as liquid CO_2 is nearly depleted. If an auto changeover is fitted, has it switched? Check the indicator window.

Check 2: Is there a gas leak in the line? Pressurise the system, walk the conduit run, and listen for hissing. Apply soapy water to all fittings, bubbles indicate a leak.

Check 3: Is the solenoid receiving a signal from the pH controller? Switch the controller to manual. Does the CO_2 controller activate? If not, check wiring and signal cable connections.

Check 4: Is the flow meter ball showing movement? If the ball is stationary with the solenoid open, check for a blockage at fittings or at the diffuser.

Water looks grey, dull, lifeless

Likely cause: High total alkalinity combined with unbalanced water. This is a classic presentation before the system has been properly commissioned. Check TA, calcium hardness, and LSI.

Action: Bring TA to target first using acid doses. Do not raise calcium hardness until TA is below the calcium hardness value. Water clarity typically improves markedly once chemistry is balanced.

Change slowly. Change one thing at a time. Allow at least one full water turnover before assessing the result. Trust the regular operator's knowledge of this specific pool.

CHAPTER 31

Glossary

CO_2: Carbon dioxide; used as a pH control agent in commercial pools.

pH: A measure of acidity or alkalinity; critical for swimmer comfort and chemical balance.

Total Alkalinity: Water's ability to buffer pH changes, expressed as mg/L as $CaCO_3$.

Calcium Hardness: Concentration of dissolved calcium; influences scaling and corrosion tendency.

LSI: Langelier Saturation Index; indicates the scaling or corrosive tendency of water.

Hypochlorite: A chlorine compound, such as sodium hypochlorite, used for disinfection.

Diffuser: A device that introduces CO_2 into water as fine bubbles to maximise absorption.

Venturi Injector: A device that draws gas into the water stream using pressure differential.

Bypass Contact Chamber: A side-stream loop used to extend CO_2 absorption contact time.

Specific Gravity: The ratio of the density of a substance to the density of water.

Equivalent Weight: The mass of a substance that supplies one mole of reactive ions.

Henry's Law: The physical law describing gas solubility in liquids based on partial pressure.

Regulator: A device controlling gas delivery pressure from CO_2 cylinders or bulk storage.

Aeration: The introduction of air into water; accelerates CO_2 loss and drives pH upward.

Turbulence: Water motion that increases gas exchange and contributes to pH rise.

PART 4 — UNDERSTANDING THE SCIENCE

CHAPTER 32

Why the Science Matters (But Stays Practical)

The appendices on LSI and Henry's Law explain why the earlier practical advice works

Read them when time allows. Understanding how gas leaves water and how calcium balance behaves will save money, reduce chemical waste, and settle arguments on the pool deck

CHAPTER 33

Appendix:

Langelier Saturation Index

Quick take: LSI balances calcium carbonate. You can satisfy the LSI and still operate at a lower TA to reduce CO_2 demand, if surfaces and local regulations allow.

Early in my career, I believed as many in the industry still do that standard pool practice was applying the Langelier Saturation Index correctly. Over time, I came to realise this was not quite the case.

Most LSI calculations used in the pool industry are derived versions of the original index. They are related to it, but they are not identical.

As originally developed, the LSI is a calcite saturation index. Its purpose is to indicate whether water is likely to dissolve calcium carbonate or deposit it as scale:

- A negative LSI indicates water capable of dissolving calcium carbonate
- A positive LSI indicates water capable of forming calcium carbonate scale
- An LSI near zero represents balance under the assumed conditions

Temperature, evaporation, and operating conditions can shift this balance over time.

Since its original publication, the LSI has been modified in various ways to better suit pool and spa applications. Suppliers, organisations, and training bodies have each introduced simplified versions suited to their own products and tools. Mobile apps and calculators have added further variations, often without clearly stating the assumptions behind them.

The result is that two different tools can calculate "LSI" for the same water and produce different answers.

This confusion has been embedded in the industry for decades. At trade nights, suggesting a different approach to water balance has occasionally prompted heated debate. The problem has never been the concept itself, it has been the assumption that all "LSI" calculations are the same.

The pool industry commonly uses a practical form of the LSI that looks like this:

SI = pH + Temperature Factor (TF) + CH Factor (CF) + TA Factor (AF) – 12.1

Temp (0C)	TF	Total Alkalinity	AF	Calcium Hardness	CF
0	0	5	0.7	5	0.3
3	0.1	25	1.4	25	1
8	0.2	50	1.7	50	1.3
12	0.3	75	1.9	75	1.5
16	0.4	100	2	100	1.6
19	0.5	150	2.2	150	1.8
24	0.6	200	2.3	200	1.9
29	0.7	300	2.5	300	2.1
32	0.8	400	2.6	400	2.3
41	0.9	800	2.9	800	2.5
51	1	1000	3	1000	2.6

This equation can be simplified further by using 24 °C as a constant temperature:

SI = pH + CH Factor (CF) + TA Factor (AF) – 11.5

This highlights a central issue with how LSI is applied in practice. Because the index is so widely used, shortcuts and simplifications are common. When values are generated by software or apps, it can be difficult to know which assumptions sit behind the number being presented.

"Note: The effect of cyanuric acid on total alkalinity is deliberately set aside in the discussion that follows. At higher CYA concentrations, particularly in residential trichloro pools, this correction becomes important and can materially affect results. That complication is real, but it is not required to make the point here."

Temperature

One of the most overlooked inputs in LSI calculations is temperature. Historically, this was often justified on the basis that water temperature changes slowly and within a narrow range. For many warm pools, this reasoning has some validity.

As temperature increases, the temperature factor in the LSI equation changes only modestly. An increase from 29 °C to 34 °C shifts the index by roughly 0.1 a small movement within an operating tolerance of ±0.5. In contrast, a change from 10 °C to 30 °C alters the index by approximately 0.3, which can be significant.

This becomes increasingly important for temperature-sensitive surfaces such as plaster. Treating temperature as negligible in those cases can lead to misleading conclusions about water balance.

Simplified tables and quick-reference charts should therefore be used with care. They are useful, but they conceal the assumptions that can materially influence the result.

Equivalent LSI values — different outcomes

One limitation of relying on LSI alone is that very different water balances can produce the same index value.

To achieve an SI of approximately zero (± 0.5), both of the following conditions are valid:

Balance 1: pH 7.5, CH 75 mg/L, TA 300 mg/L, temperature 24 °C SI =

$$7.5 + 0.6 + 1.5 + 2.5 - 12.1 = 0.0$$

Balance 2: pH 7.6, CH 800 mg/L, TA 25 mg/L, temperature 24 °C SI =

$$7.6 + 0.6 + 2.5 + 1.4 - 12.1 = 0.0$$

According to the index, both are acceptable. In practice, they behave very differently. The first is difficult to hold stable, particularly with respect to pH. The second is generally easier to control.

These examples demonstrate a key point: LSI values are not fixed targets but flexible outcomes. Water does not need to sit within a narrow, prescribed range to satisfy the intent of the index.

Materials matter more than they used to

Concerns about low total alkalinity are often rooted in historical construction practices. In the past, many pools were built with materials genuinely vulnerable to corrosion.

In the late 1980s, I was asked to repair a large residential pool constructed in the style of mid-1960s school pools, hydraulically complex and plumbed entirely in copper. In systems like this, very low total alkalinity could occasionally produce low pH events that damaged pipework. In another case, water turned green following hypochlorite dosing, a classic indicator of iron in the system, not unusual when pools were built with steel pipework.

Today, such materials are rare. A modern pool built with unprotected steel pipework would be unusual, and copper is uncommon outside of specific heat-exchange components.

Modern pools typically use inert or protected materials: plastics, fibreglass, titanium, ceramics, coated cast-iron, stainless-steel filters, epoxy finishes, and polymer-based grouts. These materials are far less sensitive to the conditions that once drove conservative alkalinity targets.

> *"(Clarification: plain plaster pools remain common in the United States for commercial and public facilities, and surface sensitivity must always be considered in those contexts.)"*

What LSI is trying to protect

The Langelier Saturation Index is intended to balance calcium carbonate in solution so that water is neither scaling nor aggressive to grout or plaster. While it is often applied to metals, its original purpose is calcium carbonate equilibrium, not blanket corrosion prediction.

This does not mean alkalinity can be ignored. Calcium hardness and alkalinity must be considered together, in the context of the materials present and the control approach being used.

One sequencing point is critical: total alkalinity must be brought below calcium hardness before calcium hardness is increased. Failing to observe this order can trigger rapid calcium carbonate precipitation, clouding the water for days, regardless of what the calculated index shows.

CHAPTER 34

Appendix:

Henrys Law

The previous chapter addressed the Langelier Saturation Index, the chemical side of water balance. Water treatment, however, is not purely chemical. It is also governed by physics, and some physical processes cannot be managed away.

One of the most important is Henry's Law, which describes how gases behave in liquids.

Henry's Law states that, at a given temperature, the amount of a gas dissolved in a liquid is proportional to the partial pressure of that gas above the liquid. In practical terms: gases will leave water until equilibrium with the surrounding air is reached.

In swimming pools, this explains several everyday observations. When water is warm, agitated, or exposed to air movement, dissolved CO_2 leaves the water more rapidly, and pH rises. Reducing turbulence and covering the water at rest helps retain CO_2 and stabilise pH.

Henry's Law is not a theoretical concept applied to pools as an afterthought. It is a fundamental physical relationship that governs how pool water behaves every day, whether or not we account for it.

Henry's Law is expressed as:

$$C = k \times P$$

Where:

- C = concentration of the gas in the liquid (mol/L)
- k = Henry's constant, specific to a gas and solvent at a given temperature
- P = partial pressure of the gas above the liquid (atm or bar)

For carbon dioxide:

$$3.4 \times 10^{-2} = 0.034 = [CO_2(aq)] / [CO_2(g)]$$

Current atmospheric CO_2 concentration is approximately 422–426 ppm, expressed as:

$$4.12 \times 10^{-4} \text{ atm}$$

At 25 °C, the equilibrium concentration of CO_2 in water is therefore:

$$4.12 \times 10^{-4} \times 0.034 = 1.4 \times 10^{-5} \text{ mol/L}$$

This represents the equilibrium concentration, not an operational target. It illustrates how readily CO_2 leaves water under normal atmospheric conditions.

Proof in the real world

Henry's Law can be demonstrated simply using a spa with an air blower:

1. Balance the water to a traditional water balance: calcium hardness 200 mg/L, total alkalinity 110 mg/L, pH adjusted to 7.2–7.6.
2. Turn off all chemical control and dosing systems.
3. Measure and record the pH.
4. Turn on the blower and, if fitted, the jet booster pump.
5. Wait 15–20 minutes.
6. Measure pH again. In practice, it will have risen.

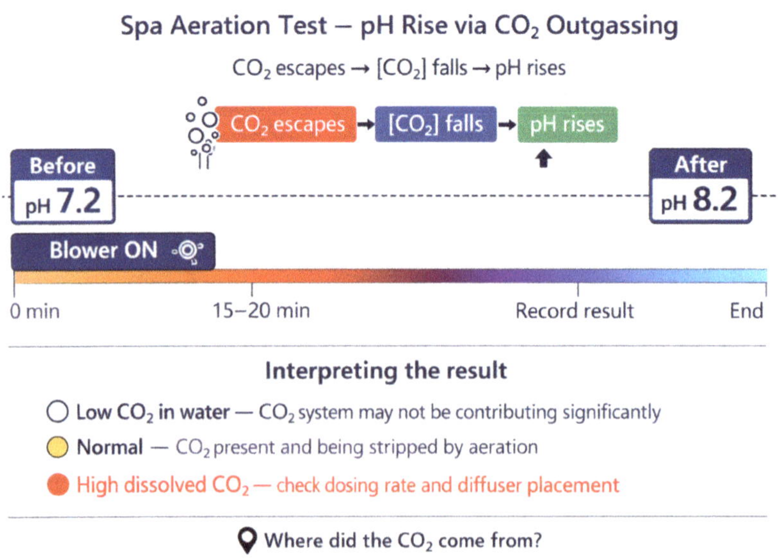

In one test conducted, a well-aerated spa showed pH rise from 7.5 to above 8.2 in approximately 15 minutes.

This effect occurs in all water bodies. The magnitude of the rise will vary with blower aggressiveness, turbulence level, and overall water movement. The underlying cause is the same in every case: dissolved CO_2 is released to atmosphere, and pH rises as it leaves.

The same process is visible wherever water is exposed to sustained turbulence or aeration:

- Fountains and sprays
- Weir edges
- Spillovers and cascades
- Any location where surface movement accelerates gas exchange

This behaviour is governed by Henry's Law, gases leave water until equilibrium with the surrounding air is reached.

Where does the CO_2 come from in the first place?

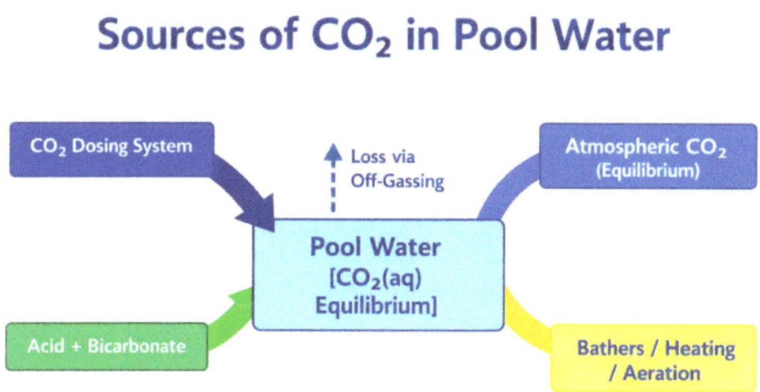

At current atmospheric CO_2 levels of approximately 400–420 ppm, air-equilibrated freshwater contains only about 0.5–0.6 mg/L of dissolved CO_2 at typical pool temperatures of 25–29 °C. This dissolved CO_2 is not alkalinity.

In freshwater at around pH 7.5, total alkalinity arises primarily from bicarbonate and carbonate species. In low-mineral water equilibrated with air, total alkalinity may be only a few milligrams per litre, approximately 2–3 mg/L as $CaCO_3$ — unless local geology contributes additional buffering.

Where CO_2 Comes From

Most dissolved inorganic carbon in pool water originates from:

- Fill water chemistry
- The carbonate buffering system

Additional CO_2 is continuously produced through:

- Acid reactions with bicarbonate
- Oxidation processes

However, the dominant factor in pool operation is CO_2 exchange with the atmosphere, not internal generation.

One useful way to estimate dissolved CO_2 in bicarbonate-buffered water is:

$$CO_2 = (TA \times 2) \times 10^{\wedge}(6.2 - pH)$$

Higher total alkalinity and lower pH both increase dissolved CO_2 concentration. This equation is not a precise control tool, it illustrates the strong dependence of CO_2 on both TA and pH.

When acid is added to bicarbonate-buffered pool water:

$$HCl + NaHCO_3 \rightarrow NaCl + H_2O + CO_2$$

Hydrochloric acid + Sodium Bicarbonate →
Salt + Water + Carbon Dioxide

Adding acid creates CO_2 in the water. That CO_2 then begins to leave the water and equilibrate with the surrounding air, in accordance with Henry's Law.

This process is continuous. The higher the total alkalinity and the lower the pH, the greater the dissolved CO_2 concentration, the faster it will outgas, and the more rapidly pH will rise.

This is why operating a pool at a TA of approximately 100 mg/L while holding pH near 7.5 using acid control produces a repeating cycle: acid addition → CO_2 generation → CO_2 outgassing → pH rise → more acid required.

In simple terms:

Acid + Bicarbonate → Carbon Dioxide + Water + Salt

Understanding Henry's Law is essential to understanding CO_2-based pH control. Pool water is not just a chemical system, it is part of its physical environment, governed simultaneously by chemistry and physics. Considering one without the other leads to incomplete and often misleading conclusions.

CHAPTER 35

Henry's Law Meets Langelier Saturation Index

Balancing LSI with atmospheric equilibrium

The preceding chapters treated the Langelier Saturation Index and Henry's Law as separate topics. This was intentional, each concept is best understood independently before their interaction is considered.

In simple domestic pools with vinyl or fibreglass surfaces, it is often possible to think about these concepts in isolation. In commercial pools, this separation breaks down.

Commercial water bodies operate under additional constraints: manufacturer warranties, bather comfort expectations, and regulatory oversight. In that environment, considering chemical balance and physical behaviour independently is not sufficient. A combined understanding is required.

Most health authorities specify an acceptable pH range of 7.2–7.8, with minor regional variation. Within the industry, a narrower range of 7.4–7.6 is commonly recommended based on chlorine activity, with 7.5 often treated as an ideal target. A pH of 7.5 is used as the representative operating point in the discussion that follows.

Returning to one of the earlier LSI examples:

pH 7.6, Calcium Hardness 800 mg/L, Total Alkalinity 25 mg/L, Temperature 24 °C SI = 7.6 + 0.6 + 2.5 + 1.4 − 12.1 = 0.0

Reconciling LSI with a practical pH

Suppose the desired operating pH is 7.5 rather than 7.6. With calcium hardness reduced to 500 mg/L, the hardness factor is approximately 2.4.

At pH 7.5 and TA 25 mg/L:

$$SI = 7.5 + 0.6 + 2.4 + 1.4 - 12.1 = -0.2$$

This indicates water that is slightly aggressive toward calcium carbonate.

Rather than increasing calcium hardness further, a more effective correction is to raise total alkalinity. With TA increased to 50 mg/L, calcium hardness held at 500 mg/L, and pH at 7.5:

$$SI = 7.5 + 0.6 + 2.4 + 1.7 - 12.1 = +0.1$$

A relatively small increase in alkalinity moves the water from a marginally aggressive condition to a slightly positive saturation index.

Total alkalinity should be reduced before increasing calcium hardness. Adding calcium while alkalinity is higher increases the risk of calcium carbonate precipitation, which can cloud the water for days regardless of what the saturation index shows.

How this is clarified when Henry's Law is considered

Natural freshwater bodies, streams and lakes, commonly operate at a near-neutral pH of approximately 7.0, with total alkalinity typically in the range of 30–40 mg/L. This is a common order of magnitude for surface water that has equilibrated with the atmosphere.

This raises an obvious question: why should a swimming pool be fundamentally different? Does the Langelier Saturation Index prevent operation at similar alkalinity levels?

At first glance, industry training and long-standing practice suggest that a total alkalinity of around 30 mg/L is too low. The standard argument is that the LSI would become negative, the water would be aggressive, and surfaces would be damaged.

However, when the assumptions behind the LSI are examined carefully, and when Henry's Law is considered alongside it, the situation is not as rigid as it appears.

Temp C	CH	TA	pH	SI
24	150	150	7.6	-0.44
12	200	150	7.8	-0.1
34	500	25	7.5	0.0
24	500	25	7.5	-0.2

All four water balances in the table satisfy the saturation index equation and fall within the commonly accepted ±0.3 (industry standard operating range) tolerance. Each is chemically reasonable.

Note: Saturation index calculations shown here assume zero cyanuric acid. Where CYA is present, alkalinity must be corrected accordingly.

In practice, however, only one of these balances produces a pH that moves slowly, the case with total alkalinity around 25 mg/L. While a slightly higher alkalinity is generally preferred, this example clearly illustrates the principle.

Where local regulations specify a minimum total alkalinity, those requirements take precedence. Any departure from mandated targets must be formally documented and approved.

What these examples demonstrate is that it is possible to satisfy the intent of the Langelier Saturation Index and, at the same time, operate in a way that aligns with Henry's Law. The two are not in conflict. When applied together, they allow water balance to be adjusted deliberately, reducing pH instability and preventing surface damage, while still meeting industry and regulatory expectations.

PART 5 — TECHNICAL APPENDICES & TOOLS

CHAPTER 36

Quick Reference and Extras

Toolbox, maths, painting, checklists; note variations

This section is the toolbox, worked mathematics, conduit painting guidance, checklists, and start-up steps that can be photocopied for site use.

The aim is repeatable results. If a specific site differs from the assumptions used, note the variation next to the relevant checklist item and continue.

CHAPTER 37

Appendix:

Calculating Tube Size Example

This appendix is optional and operates at a different level of detail from the rest of the book.

The calculations are advanced and intended for experienced installers. If you are new to the industry, primarily an operator, or simply not interested in the detail, this section can be safely skipped. It is included as a reference, not as required reading.

The appendix works through the mathematics of how a longer or narrower tube restricts gas flow, and how a larger internal diameter or shorter run delivers more stable gas at the same pressure.

Do not worry about following every step in detail. In the pool industry, we often say "it's not rocket science" but in this case, the underlying equations are the same compressible-flow relationships used in rocket nozzle design. The intent here is practical understanding, not mathematical mastery.

Use this appendix only if:

- The CO_2 system is underperforming
- Tube length or diameter is unusual for the site
- A design needs to be verified beyond what the standard tables provide

Otherwise, use the sizing table in Chapter 19: Tube and Fittings.

Assumptions used for the worked examples:

- Room temperature is steady (approximately 20–25 °C)
- Straight, horizontal tube run (approximately 30 m)
- Smooth plastic tube (nylon/PU/PE)
- Outlet is near atmospheric pressure (suction side)
- Regulator set to approximately 200–300 kPa gauge (add ~101 kPa to obtain absolute pressure)

Roadmap: 9 Steps to Estimate CO_2 Tube Flow

Gas Need (Steps 1–3)

Known values
Pool volume, Cl_2 feed rate, diffuser type

CO_2 chemical need
Chlorine feed rate
✖ 0.88 rule of thumb

Add stripping losses
Feature type • loss factor
✔↑ total CO_2 / day (kg)

Pipe Geometry (Steps 4–6)

Gas flow rate
Convert kg/day ↔ L/min at delivery pressure

Measure run length
Pipe length from regulator to diffuser

Add fittings
Count elbows, tees, valves, length (FIG-06)

Tube Selection (Steps 7–9)

Select tube OD
Nearest standard size with ID>> minimum

Verify and record
Re-check with installed equiv. length

Output: Minimum tube ID confirmed
"Select next standard size up for safety margin"

Long runs, high flow, or low temperatures require full enginering calculation. Consult supplier's sizing tool.

Step 1: Area from tube size

Use the inside diameter (ID). Convert mm to metres, then calculate cross-sectional area using:

$$A = (\pi \times D^2) / 4$$

Example IDs:

- 6×4 → D = 4 mm
- 8×5 → D = 5 mm
- 10×6.5 → D = 6.5 mm

To convert to metres, divide by 1,000. For example:

$$4 \text{ mm} \div 1{,}000 = 0.004 \text{ m}$$

$$A = (\pi \times 0.004^2) / 4 \ A = (\pi \times 0.000016) / 4 \ A = \pi \times 0.000004 \ A = 0.000012566370614 \text{ m}^2$$

For the worked example, 8×5 mm tube is used (ID = 5 mm):

$$5 \text{ mm} \div 1{,}000 = 0.005 \text{ m}$$

$$A = (\pi \times 0.005^2) / 4 \ A = (\pi \times 0.000025) / 4 \ A = 0.0000785398163375 / 4 \ A = 0.000019634954084375 \text{ m}^2$$

Step 2: Pressures to use

Convert the regulator setting from kPa gauge to Pa absolute by adding atmospheric pressure (approximately 101,325 Pa).

Example: 250 kPa gauge →
250,000 Pa + 101,325 Pa → $P_1 \approx 351{,}325$ Pa

For the discharge side, a diffuser is required. Diffuser types are covered in Chapter 22 (Injection Point). The table below outlines typical diffusers and the pressures required to drive them.

Diffuser / Injector	Gas side Pressure Required at Diffuser (typical) bar	Gas side Pressure Required at Diffuser (typical) kPa	Pressure Required at Outlet (With atmospheric added) Pa
Ceramic micro-bubble LP grade	0.3-1.0	30-100	201,325
Ceramic micro-bubble HP grade	1.0-2.0	100-200	301,325
Venturi Injector	~20–30% ΔP	~20–30% ΔP	~20–30% ΔP
Membrane dis/tube (EPDM/PTFE-coated EPDM)	0.1-0.3	10-30	131,325
Sintered metal sparger(316L)	0.2-2.0	20-200	301,325

For this example, a ceramic microbubble LP grade diffuser is assumed:

$$P_2 \approx 201{,}325 \text{ Pa}$$

"Note: If completing the equations with a venturi injector, use one of the lower pressure values from the table unless the delta pressure is known."

Step 3: Mean density

The density of the gas inside the tube needs to be known — that is, how tightly packed the CO_2 is. Calculate it once at inlet pressure and once at outlet pressure, then take a simple average.

Values required:

- P_1 = inlet absolute pressure (Pa) from Step 2: 351,325 Pa
- P_2 = outlet absolute pressure (Pa) approximately 201,325 Pa
- T = temperature in Kelvin, room temperature 22 °C → 295.15 K (add 273.15)
- R_s (for CO_2) = 188.9 J/kg·K (treat as a fixed constant)

Formula: $\rho = P / (R_S \times T)$.

"Note: To convert any temperature to Kelvin, add 273.15 to the Celsius value. Using this example: 22 °C + 273.15 = ***295.15 K"***

Formula:

density = pressure ÷ (R_S × temperature)

Written mathematically as:

$\rho = P / (R_S \times T)$.

Worked example: outlet density (ρ_2)

The outlet pressure is assumed to be 0 kPa gauge, as the injection point is on the suction side of the pump. The pressure could be slightly negative in practice, but 0 kPa gauge is used here for simplicity.

At 0 kPa gauge, we are at one atmosphere, which equals approximately 101,325 Pa absolute. As established in Step 2, P_2 for the LP ceramic diffuser is approximately 201,325 Pa absolute.

201,325 ÷ 55,800 ≈ **3.6 kg/m³**

$\rho_2 \approx 3.6$ kg/m³

Mean density (the value used in subsequent steps)

Average the two results:

$$\rho m = (\rho_1 + \rho_2) \div 2$$

$$\rho m = (6.3 + 3.6) \div 2 \approx \mathbf{4.96\ kg/m^3}$$

What the numbers are showing:

- At the high-pressure inlet, CO_2 is more tightly packed (~6.3 kg/m^3)
- At the outlet, it is looser (~3.6 kg/m^3)
- Using the average of approximately 5.0 kg/m^3 is sufficient for this application

Quick reference: mean density by regulator setting

(T ≈ 295 K, approximately 22 °C)

Regulator kPa at gauge	P_1 absolute Pa	ρ_1 (inlet) kg/m^3	ρ_2 (outlet) kg/m^3	ρm to use
200	301,325	≈ 5.41	≈ 3.61	≈ 4.51
250	351,325	≈ 6.30	≈ 3.61	≈ 4.96
300	401,325	≈ 7.20	≈ 3.61	≈ 5.41

Step 4: Friction factor f

For smooth plastic tube, a field estimate is appropriate. Start with:

$$f = 0.02$$

This is a safe initial value. It will be updated after Step 6 using the Blasius equation:

$$f \approx 0.3164 \times Re^{-0.25}$$

Example with Re = 34,000:

$$Re^{-0.25} = 1 / Re^{0.25}\ Re^{0.25} = 34{,}000^{0.25} \approx 14.5\ Re^{-0.25} \approx 1 / 14.5 \approx 0.069\ f \approx 0.3164 \times 0.069 \approx 0.022$$

Note: Reynolds number uses the velocity calculated in Step 6. Run Step 4 initially with f = 0.02, then return with the Blasius result. In practice, this typically lands between 0.021 and 0.026.

Step 5: Pressure push term

Calculate $(P_1^2 - P_2^2)$. This captures the effect of gas compressibility without complex mathematics.

$$P_1 = 351{,}325 \text{ Pa} \quad P_2 = 201{,}325 \text{ Pa}$$

$$351{,}325^2 - 201{,}325^2 = 123{,}429{,}255{,}625 - 40{,}531{,}755{,}625 = \mathbf{82{,}897{,}500{,}000\ Pa^2}$$

Step 6: Mass flow (Corrected Isothermal Compressible Form)

Formula:

$$\dot{m} = (A/2) \times \sqrt{(D / (f \times L))} \times \sqrt{((P_1^2 - P_2^2) / (R_S \times T))}$$

Where:

- A = cross-sectional area (m^2)
- D = internal diameter (m)
- f = Darcy friction factor
- L = pipe length (m)
- P_1, P_2 = upstream and downstream absolute pressures (Pa)
- R_S = specific gas constant ($J \cdot kg^{-1} \cdot K^{-1}$)
- T = absolute temperature (K)

Values carried forward:

- D = 0.005 m
- A = 0.000019634954084375 m^2
- $(P_1^2 - P_2^2)$ = 82,897,500,000 Pa^2
- f = 0.022
- L = 30 m
- R_S = 188.92 $J \cdot kg^{-1} \cdot K^{-1}$
- T = 295.15 K

Working through the components:

$$A/2 = 0.000019634954084375 \div 2 = 0.0000098174770421875\ m^2$$

$$f \times L = 0.022 \times 30 = 0.66$$

$$D / (f \times L) = 0.005 \div 0.66 \approx 0.007576$$

$$\sqrt{(D / (f \times L))} \approx \sqrt{0.007576} \approx \mathbf{0.087028}$$

$$\sqrt{(P_1^2 - P_2^2)} = \sqrt{82{,}897{,}500{,}000} \approx \mathbf{287{,}919.26\ Pa}$$

$$R_S \times T = 188.92 \times 295.15 = 55{,}759.738$$

$$1 / \sqrt{(R_S \times T)} = 1 / \sqrt{55{,}759.738} = 1 / 236.135 \approx 0.004235$$

Assembling the result:

$$\dot{m} = 0.0000098174770421875 \times 0.087028 \times 287{,}919.26 \times 0.004235$$

$$\dot{m} \approx \mathbf{0.001042\ kg/s}$$

Step 7: Velocity

Values carried forward:

- $A = 0.000019634954084375\ m^2$
- $\rho m = 4.958681394\ kg/m^3$
- $\dot{m} = 0.001042\ kg/s$

$$\text{Formula: } v = \dot{m} \div (\rho m \times A)$$

$$\rho m \times A = 4.958681394 \times 0.000019634954084375 = 0.00009736\ m^2 \cdot kg/m^3$$

$$v = 0.001042 \div 0.00009736 \approx 10.70\ m/s$$

Step 8: Optional Mach number check

Speed of sound in CO_2 at room temperature:

$$a = \sqrt{(\gamma \times R_S \times T)}$$

$$\text{For } CO_2\text{: } \gamma \approx 1.30$$

$$R_S \times T = 55{,}759.738 \text{ (from Step 6)}$$

$$a = \sqrt{(1.30 \times 55{,}759.738)} = \sqrt{72{,}487.66} \approx 269.23\ m/s$$

$$\text{Mach} = v \div a = 10.70 \div 269.23 \approx 0.040$$

A Mach number well below 0.3 confirms the flow is safely subsonic and this calculation method is valid.

Step 9: Standard volumetric flow (for datasheet comparison)

Equipment manufacturers publish flow ratings under standardised conditions, not site conditions. To compare results directly, convert mass flow to standard volumetric flow:

$$\dot{V}\_std = \dot{m} \div \rho\_std$$

A. Standard conditions at 20 °C, 1 atm ($\rho\_std = 1.80\ kg/m^3$):

$$\dot{V}\_std = 0.001042 \div 1.80 = 0.000579\ m^3/s$$

= 0.579 L/s = 34.73 SL/min (standard litres per minute at 20 °C)

B. Normal conditions at 0 °C, 1 atm ($\rho\_std = 1.98\ kg/m^3$):

$$\dot{V}\_N = 0.001042 \div 1.98 = 0.000526\ m^3/s$$

$$= 0.000526 \times 3{,}600 = 1.894\ Nm^3/h$$

Temperature °C	Approximate rho_std for CO_2 kg/m3
0	1.96
15	1.86
20	1.83
25	1.80

CHAPTER 38

Appendix

Painting Plastics

Painting plastic is a frustrating task. Paint goes on well, looks fine, dries, and then peels away the moment the surface is touched or moved.

This happens because plastic surfaces carry a layer of oils known as plasticisers. These prevent paint from bonding directly to the plastic. Instead, the paint bonds to the plasticiser layer, which is effectively an oily coating, and the result is that paint is bonding to nothing that will hold it.

To paint plastic successfully, the plasticisers must be removed first.

There are three approaches:

1. **Heat:** Plastics manufacturers use flaming prior to printing to drive plasticisers off the surface. A heat gun can achieve the same result. The technique is to heat the surface sufficiently to allow the plasticisers to volatilise. Paint should be applied while the surface is still warm.

2. **Special undercoats:** Hardware stores stock undercoats specifically formulated to prime plastic. These may be more expensive than other options but can be the most practical solution depending on location, local regulations, and safety requirements. As with any chemical used on site, obtain, retain, and follow the Safety Data Sheet.

3. **Chemical treatment:** Solvents can strip plasticisers and prepare the surface for painting.

How to Paint Plastic Conduit (3 Valid Methods)

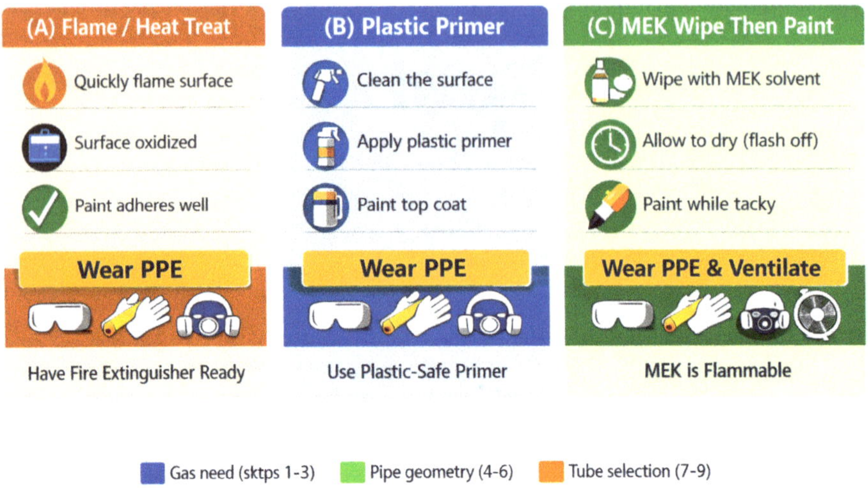

The current method used in our installations is methyl ethyl ketone (MEK). It is essential to obtain, read, and retain the Safety Data Sheet for this product before use. MEK should be used outdoors or in a well-ventilated area, with all PPE specified in the SDS.

In practice, the conduit is wiped with MEK and painted immediately while the surface is still tacky. MEK is a component of PVC glue, when applied, it causes the plastic surface to become slightly sticky, and the paint is applied directly to that prepared surface. The result is that the paint is effectively bonded to the plastic rather than sitting on top of it. Once dry, the finish is substantially more resistant to scuffs and abrasion. A second coat over the first will typically last years rather than days.

For small-bore conduit, painting before fixing it in place is strongly recommended. Painting conduit once it is mounted is possible but fiddly.

CHAPTER 39

Appendix

Installation and Commissioning Checklist

Date	
Client	
Client Contact	
Site	
Waterbody	
Volume	
Temperature	

Water Tests

	Fill Water	Before Start	Start Up	Day 1	Day 2	Day 3	Day 4
Free Chlorine							
Total Chlorine							
pH							
Calcium Hardness							
Total Alkalinity							
Cyanuric Acid							
Total Dissolved Solids							

Record Settings

Use or add additional lines to suit the site.

	First Visit	Start Up	Day 1	Day 3
pH Set point				
Chlorine Set point				
CO_2 regulator pressure				
CO_2 Flow rate				

CHAPTER 40

Appendix:

Start Up Steps

Step 1: Do Nothing (Deliberately)

Once the installation is complete and leak-tested:

- Enable CO_2 control
- Set the initial flow conservatively, use the manufacturer's recommendation or a mid-range setting
- Do not alter controller settings immediately
- Record all settings in the form provided

Then stop and observe.

Watch for:

- How often the controller calls for correction
- Whether pH is drifting slowly or sharply
- Whether bather activity visibly affects the reading
- Whether chlorine is feeding continuously or intermittently
- Whether undissolved bubbles are visible at the pool returns

Observe for at least 20–30 minutes. Longer if time allows.

The goal at this stage is not to fix. It is to understand.

Step 2: Speak to the Regular Operator

The regular operator knows this water body better than anyone else. This is not necessarily the centre manager or duty manager, it is the person who routinely adjusts the chemical controller.

Ask:

- Is today's bather load typical?
- Is this chlorine feed rate normal?
- Does pH usually move this way?
- Are we in peak or quiet season?
- Has anything changed recently?

Do not override their experience with theory. Field knowledge outweighs first-day assumptions.

Step 3: Record, Don't React

Record:

- pH set point
- Control band
- Proportional / integral values (if adjustable)
- Chlorine set point
- Current CO_2 regulator pressure

These values are the baseline. Do not change them unless behaviour is clearly unstable.

Change slowly. Change one thing at a time. Change incrementally. Wait at least one full water turnover before assessing the result.

Step 4: Allow the System to Find Its Balance

CO_2 control is softer than acid control. Allow time for:

- Dissolution equilibrium to establish
- Chlorine demand patterns to emerge
- Aeration effects to become apparent
- Normal daily cycles to play out

If pH is within regulatory limits and moving predictably, leave it alone. Small drift is not failure.

Step 5: Return the Next Day

The real commissioning check happens 24 hours later. Return at approximately the same time of day.

Check:

1. Has pH stabilised within the control band?
2. Has total alkalinity risen noticeably?
3. Has CO_2 consumption been excessive?
4. Has the controller short-cycled?

Only now, if required, make minor adjustments.

> *"Reminder: Change slowly. Change one thing at a time. Change incrementally. Wait at least one full water turnover before assessing the result."*

Guiding Principles

- Act intentionally
- Adjust slowly
- Change one variable at a time
- Trust observed behaviour over assumption
- Defer to the operator's lived experience
- Never tune a system you do not yet understand

This is a guide, not a fixed protocol. Some sites will require daily visits for three to four days; others may need support for a week or more. Be generous with time. This process is not infallible.

> *"Note: Local health codes and the requirements of the Authority Having Jurisdiction (AHJ) supersede this guidance. Always document site-specific compliance."*

CHAPTER 41

Appendix:

Global Standards Crosswalk for CO_2 pH Control

This section provides a comparative overview of international standards governing the use of carbon dioxide for pH correction in commercial aquatic facilities. It indicates whether CO_2 is permitted, outlines key requirements, and includes notes relevant to operators and regulators.

Region / Standard	CO_2 Allowed?	Key Requirements	Notes
North America / PHTA/ICC (ANSI/APSP/ ICC-1 & -2)	Yes	CO_2 permitted as pH control agent; must comply with safety, ventilation, and injector placement guidelines.	Widely used in commercial pools; referenced in CPO® Handbook. Local codes may vary.
Europe / BS EN 15288-1:2018 + A1:2024	Yes (implicitly)	Focuses on safe design and operation; CO_2 not explicitly named but allowed under chemical dosing provisions.	CO_2 use must meet chemical handling and safety standards. Inspector discretion applies.
Germany / DIN 19643-1:2023	Yes	CO_2 recognized for pH control; dosing must be precise and integrated with water treatment systems.	Highly technical; emphasizes process control and water quality outcomes.
Australia/ NZ / AS/NZS 1838:2021, 1839:2021, 2783:2025	Yes	CO_2 allowed; must meet gas safety, injector design, and water balance requirements.	Local health departments may require additional documentation or approval.

Health Inspector Talking Points

The following points may be useful when discussing CO_2 pH control systems with health inspectors or regulatory authorities:

- CO_2 is recognised in multiple international standards as a safe and effective pH control method.
- Correct injector placement and gas dissolution techniques minimise off-gassing and support stable pH.
- CO_2 systems reduce chemical handling risks compared to mineral acids, improving operator safety.
- Water balance can be maintained within LSI guidelines using CO_2 with appropriately adjusted alkalinity and calcium hardness.
- CO_2 dosing systems can be integrated with automated controllers for precise and consistent regulation.
- Facilities using CO_2 have demonstrated reduced chemical consumption and improved water quality outcomes.
- Documentation of injector setup, safety protocols, and water balance calculations supports regulatory compliance.

CHAPTER 42

About the Author

David Watson

David Watson entered the aquatic industry by accident, quite literally. Early for a job interview at a carpet factory in Auckland, he parked outside a closed pool shop to read a book. The shop had just been broken into by an ex-business partner, and when the owner returned from the Takapuna police station, he assumed David was connected to the incident. Six and a half hours later, after a conversation that wandered from chemistry to pumps to life in general, David walked out not with a police report, but with the job offer that would shape the next forty years of his career.

His early education in water chemistry came from an unlikely mix of brilliant mentors: private tuition from Ty Webb, guidance from a Henderson Environmental Health Officer, and hands-on training from Ken, "the poison dwarf" of EziClor Chemicals, whose own lineage traced back to the Blue Haven School in Los Angeles. This combination of practical experience and scientific curiosity established David's lifelong habit of questioning assumptions, testing claims, and replacing industry folklore with real, measurable understanding.

After relocating to Western Australia, David spent over three decades with Shenton Aquatics, becoming one of the state's most experienced specialists in recirculating aquatic systems. His work spans microbial risk mitigation, hydraulic troubleshooting, regulatory compliance, and technical innovation, including serving as joint inventor on a US-filed water chemistry dosing process. Known for translating complex chemistry and physics into practical, field-ready guidance, David's approach has always been grounded in one belief: a pool is not a static container of water but a dynamic system with safety as its first obligation.

Across his career, David has worked alongside some of the most capable pool technicians in the trade, people who can rebuild a pump with a shifter and colourful language, ride motorbikes on the weekend, and outlast him at the bar without breaking a sweat. David is, by contrast, bookish, early-to-bed, happiest when working through a technical puzzle, and quietly driven by a desire to raise industry professionalism through clear thinking and better understanding.

This book continues that mission. It is his attempt to bring clarity, science, and real-world practicality to a topic that is too often misunderstood, oversimplified, or dismissed, CO_2 pH control. What follows is not theory for theory's sake. It is the distilled outcome of decades spent watching what actually happens in the water: on site, at 3 pm on a summer's day, when the toddler pool is erupting like a geyser and the pH refuses to hold still.

It is a guide written for the people who do the work, from someone who still thinks like a technician first.

Contact and Support

For further assistance, technical support, or feedback on this guide:

Website: **www.davidwatsonpoolconsulting.com**

Available at the website:

- Additional resources and downloadable forms
- Contact details for consulting and training services
- Updates and announcements

For direct support, use the contact options provided on the website to reach David Watson Pool Consulting.

www.ingramcontent.com/pod-product-compliance
Lightning Source LLC
LaVergne TN
LVHW050913210926
848482LV00012B/423
* 9 7 8 1 7 6 4 5 2 9 6 6 2 *